D0477456

THE SILENT LANDSCAPE

Also by Richard Corfield

Architects of Eternity

THE SILENT LANDSCAPE

In the Wake of HMS *Challenger* 1872–1876

Richard Corfield

John Murray

First published in Great Britain in 2004
by John Murray (Publishers)
A division of Hodder Headline

2 4 6 8 10 9 7 5 3 1

A CIP catalogue record for this title is available from the British Library

ISBN 0 7195 6530 8

Typeset in Bembo
Printed and bound in Great Britain by Clays Ltd, St Ives, plc.

John Murray (Publishers)
338 Euston Road
London
NW1 3BH

For Jessica

Contents

Frontispiece	x
Prologue	xi
Cast of Characters	xvi
Threshold of the Deep	1
The Desert Under the Sea	25
The Restless Earth	45
Kingdoms of Mud and Lime	61
Climate Triggers and Bermudan Secrets	75
Kelp and Cold Light	100
The Library of Time	127
The Grim Latitudes	147
The Lost World	160
The Echoes of Evolution	176
The Groaning Planet	195
Dreams of Big Science	222
Epilogue	249
Acknowledgments	255
Further Reading	257
Index	261

Illustrations

Frontispiece

The Route of HMS *Challenger*   x

Cast of Characters

Henry Moseley   xvi
William J. J. Spry   xvi
Officers group   xvi
Captain George S. Nares   xvii
John J. Wild   xvii
John Murray   xvii

Figures

1: Plan of *Challenger*   8–9
2: Instrument and sampling platform   11
3: Naturalist's lab   12
4: Sounding and dredging apparatus   49
5: Distribution of calcareous sediments   64
6: *Challenger* at St. Paul's Rocks   117
7: Deep-sea fauna   136–137
8: Geological timescale   139
9: *Hallucigenia* by Simon Conway Morris   190

"What a situation to be in!" I exclaimed. "To overrun these deep regions where man has never trod! Look, Captain, look at these magnificent rocks, these uninhabited grottoes, these lowest receptacles of the globe, where life is no longer possible! What unknown sights are here!"

Jules Verne, *20,000 Leagues Under the Sea*

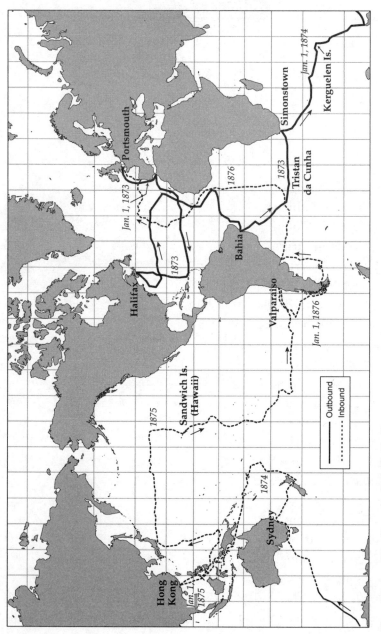

FRONTISPIECE The Route of HMS *Challenger*

Prologue

In the early spring of 1990 I flew all the way around the world.

"There's nothing remarkable about that," I hear you say, "in this day and age lots of people fly around the world." True enough, but what made my trip remarkable was that it took me two months despite being carried by jet airliners almost all the way. In this age of high technology, my trip took almost as long as Phileas Fogg's in Jules Verne's epic *Around the World in Eighty Days*, and he had been borne by ship, train, balloon, and camel. The reason for the long delay in my returning to Britain was that I had spent two months at sea in the southwestern Pacific Ocean. I joined the JOIDES (Joint Oceanographic Institutions for Deep Earth Sampling) *Resolution*, the drilling ship of the Ocean Drilling Program (ODP) in Guam and then set out on a two-month journey of undersea scientific discovery. Our brief was to drill into the seabed in the vicinity of the Caroline Islands, penetrating the submerged yet massive topographic high known as the Ontong Java Plateau, an area larger than New York State.

For two months I labored at a microscope, identifying tiny fossils retrieved by the drill cores from the ocean floor 2 miles beneath us, working 12-hour shifts, 7 days a week, for 63 days without respite. It was hard work, I was seasick and there was no privacy. Also there was the dispiriting knowledge that once the ship left port there was no way off save in the event of a life-threatening emergency, in which case the million-dollar drill string would be abandoned like a discarded syringe and the ship would sail under full power for the nearest point a helicopter could reach us. All the time the ship was at sea, it cost the worldwide consortium that ran the ODP $2,000 an hour just to keep it operating, so you can understand that they would not abandon their scientific objectives without good reason. Knowing that I was effectively a prisoner aboard a mobile drilling rig, literally on the other side of the world from friends and family, I count among the hardest things I have had to endure in my entire life.

When I got off the ship I remember quite clearly kneeling down on the quayside among the massive containers in the container port at Agana in Guam, and kissing the hot concrete in the tropical sun. I felt the freedom of the condemned man released, the sun shone brighter than I had ever seen it, the vegetation looked greener than I had thought possible, and the ground felt so *good*; so reassuringly firm and solid. I was never so pleased to be off a ship in my life and within 48 hours I was home again, safe and sound in Oxford. Yet that voyage of the ODP, like almost *all* its voyages—and those of its predecessor the Deep Sea Drilling Project (DSDP)— was hugely successful, adding immeasurably to our knowledge of the way the deep ocean and the seafloor operates. From the voyages of the ODP and the DSDP we now know about the intricacies of seafloor formation, the way that the deepwater in the ocean circulates and controls the world's weather and climate, the location of vital energy reserves as well as the places where new forms of life— and new medicines—are to be found.

My discomfort had actually contributed some good, added some tiny morsel to the sum total in the human knowledge database. In the intervening years I have become proud of what I endured and achieved. Yet with that pride has come the knowledge that my 63 days at sea were as nothing compared to the hardships endured by the marine scientists of another age. Darwin's voyage aboard *Beagle* lasted five years after all, and Huxley's aboard

Rattlesnake not much less. Yet there is another voyage of the nine-teenth century—indeed it was the last such voyage of the Victorian era—that is not so well known, and it single-handedly founded the sciences that we today know as oceanography and marine geology.

That was the voyage of HMS *Challenger*.

The *Challenger* expedition was a scientific circumnavigation of the world that lasted almost four years and traversed 68,900 miles. *Challenger* left Portsmouth, England, in December 1872 and returned in May 1876, having traveled as far as the Great Ice Barrier of Antarctica, visiting Nova Scotia, the Caribbean, and South Africa in the process, before pushing on into the Pacific, visiting Indonesia and passing not far from the Caroline Islands where I would one day undergo my own personal and scientific voyage of discovery. From the southwestern Pacific, *Challenger* headed north to Hawaii, then south again before passing back into the Atlantic through the narrow straits at Tierra del Fuego. The homeward stretch took her up through the Atlantic, into the Channel and then, finally, home again. In the course of its epic voyage fully a quarter of *Challenger*'s crew complement of 269 deserted, distressed by confinement in a ship that was only 200 feet long and 40 feet wide and demoralized by the endless repetitive grind of dredging the seabed and retriev-ing what looked to the untutored eye like lumps of mud.

Despite the stresses and strains of *Challenger*'s epic voyage, the result was a resounding success. The scientific report eventually ran to 50 volumes and took 20 years to complete. A copy of it is deposited in the Bodleian library of the University of Oxford and it was a chance encounter with it in the stack room there that first gave me the idea for this book. But perhaps even more fascinating than the official 50-volume report are the diaries kept by members of *Challenger*'s crew. It is from these particularly that I have drawn the narrative of *The Silent Landscape*. Several crewmen and scientists wrote movingly about their experiences aboard *Challenger* during those four years: the terrestrial naturalist Henry Moseley, whose interest in the lands they visited and vexation with the boredom of continual dredging was so obvious that he relegated discussion of the dredging to a single chapter at the end of his book; Engineering Sub-lieutenant William J. Spry, whose detailed observations of land, wind, and sea breathe life into Moseley's scientific narratives; Navigating Sub-lieutenant Herbert Swire, whose irreverent take on the life of the upper decks pricked the pomposity of the scientists; Lord George Campbell, whose own diaries illuminate life at sea from the perspective of a member of the British aristocracy; and finally, and perhaps most extraordinarily of all, the only surviving account (indeed perhaps the only account ever written) of life below decks: the diary of Joseph Matkin, ship's steward's assistant. Only Matkin wrote about the downside of life aboard *Challenger*, the tensions that were the plight of so many men in close proximity and the effort it took to stay interested and alert for four long years in a science that only a handful on board were educated enough to understand. These accounts, together with the *Challenger*'s 50-volume report, form the narrative backbone of *The Silent Landscape*.

But I could not be satisfied with writing only a historical account of the *Challenger* expedition, because if there is one lesson that science teaches us, it is that it stands still for nobody. It moves forward inexorably and with increasing rapidity. So *The Silent Landscape* is also the science of the *Challenger* expedition updated, focusing not just on what the expedition did find but what it *would have found* if it had on board someone with a knowledge of early twenty-first-century biology, physics, and chemistry. And that person, suitably helped with up-to-date accounts of modern oceanographic and marine science, will, I hope, be you.

So prepare to join Henry Moseley, William Spry, Herbert Swire, Joseph Matkin, and the others on board HMS *Challenger* on their epochal journey around the world. Only you have the 20:20 scientific hindsight to fully appreciate what they found.

Have a good trip.

Richard Corfield
Oxford, 2003

THE SILENT LANDSCAPE

Henry Moseley. Adventurous yet sensitive, an expedition scientist whose interests lay more with the land than the sea!

William J. J. Spry. Engineering sub-lieutenant and unlikely author of the Victorian bestseller *The Cruise of HMS* Challenger.

Charles Wyville Thomson (*in white*) flanked by the bearded and charismatic Lord George Campbell (*left*) and the young raconteur Herbert Swire (*right*).

Captain George S. Nares. First captain of the *Challenger* expedition and later to become a distinguished Arctic explorer.

John J. Wild. The expedition artist and the man responsible for the many beautiful drawings of sea-creatures that graced the expedition's report.

John Murray. Fiery-tempered yet a brilliant oceanographer, Murray financed the bulk of the expedition's 50-volume report from a fortune he amassed as a direct result of the voyage.

Threshold of the Deep

Portsmouth, England, December 20, 1872, 50° 48′ N, 1° 05′ W

THE ABYSS OF TIME

If you could stand on the brow of the hill behind the English town of Portsmouth in late December of the year 1872, you would see below you a harbor crowded with warships. To the left—the Portsea Island side—are the new men o' war; three-decked iron-clad, steam powered battleships like HMS *Warrior*, as well as other, older, square-rigged, wooden men o' war little changed since the time of Nelson. These are the ships that show the flag in the farthest-flung corners of empire. It is invigorating to see these potent symbols of dominion ready to put to sea at an hour's notice, able to bring the might of the world's greatest empire to any situation that might need it. To the right—the Gosport side—are the government offices, dry docks, cranes, piers, and jetties. On this side, too, are the troop transports and official yachts. In the middle the open water of the fairway is crowded with pinnaces, jolly-boats, cutters, and pleasure steamers. The total effect is of controlled bustle and purpose in this nerve center of the Victorian navy, the clearinghouse for the most potent military symbol in the world.

But what's that in the corner? Almost hidden by the gunmetal-gray flanks of the huge battleships are the three small masts of an unremarkable corvette, a ship tiny in comparison with the hulking leviathans around it, but one that is about to change the face of science forever: HMS *Challenger*.

The purpose of HMS *Challenger*'s voyage was primarily scientific; her orders were to explore the ocean and the ocean floor not just in the waters around Britain or even her imperial territories, but also across the entire world, or at least as much of it as could be done in the three-and-a-half years allotted to the expedition. Encouraged by the successes of the recent scientific expeditions that had been sent out with the express purpose of studying the sea and its inhabitants, the Admiralty had authorized the voyage only the year before.

This sudden naval interest in the seafloor was a direct consequence of the enthusiasm and activities of two scientists: Charles Wyville Thomson and William Carpenter. Wyville Thomson was professor of natural history at the University of Edinburgh, an institution that, since the early years of the nineteenth century, had a long and distinguished interest in the natural sciences. The University's first professor of natural history, Robert Jameson, had trained some of the most notable natural historians of the nineteenth century; his most famous alumnus was Charles Darwin. Edward Forbes, another Jameson-trained naturalist, eventually succeeded him in the professorship. Although Forbes occupied the Chair for only a few months, he successfully seeded one of the strangest ideas of the mid-nineteenth century, namely that below 300 fathoms (1,800 feet) no life could exist in the ocean. This was the so-called "azoic" (a = without, zoic = life) theory and the supposed lifeless region below 300 fathoms was the "azoic zone," an abyss where no life could exist.

The professorship of natural history in the University of Edinburgh was far and away the most influential position in natural sciences in the Victorian Empire—if not the world—and the pronouncements of the post-holder tended to be accepted as received wisdom by a Victorian society obsessed with science. Resonating with the typical Victorian's preoccupation with death, the azoic theory assumed fundamental importance in the public imagination as well as the scientific mind. But Wyville Thomson was not convinced by the azoic theory. He had been present when the Norwegians dredged clear evidence of organic remains from a fjord known to be more than half a mile deep, and he had heard from his friend Fleeming Jenkin, Edinburgh's first professor of engineering, that the frayed ends of a broken telegraph cable recovered from the depths of the Mediterranean had barnacles encrusting it. To Wyville Thomson, Forbes's azoic theory could not be correct.

Wyville Thomson was determined to investigate Forbes's theory scientifically and systematically. In this endeavor he was extremely fortunate to have friends in high places, in fact, one particular friend in one of the highest places in all of Victorian science—his friend William Carpenter was a vice-president of the Royal Society. Senior officers of the "Royal," then as now, occupied a strange but enormously powerful position in the scientific world because they were both bureaucrats and scientists. On the one hand, they were capable of steering, if not actively diverting, funds into a cause that they were interested in championing and on the other, they were scientists enough to know that in so doing they were not about to make fools of themselves.

Carpenter had worked with Wyville Thomson in the 1860s and both shared a common love of the lower invertebrates—single-celled foraminifera, echinoderms such as starfish, and the sponges whose place in the grand scheme of life, then as now, was far from certain. Carpenter was persuaded by his friend that the azoic theory was worth investigating and so Carpenter in turn persuaded the Admiralty to let Wyville Thomson have the use of the steam frigate HMS *Lightning* for part of the summer of 1868. The ship sailed between the Faeroes and the Shetlands throughout that wet and windy summer, and despite cramped conditions on board and the inadequate resources of a boat that had not been fitted out for scientific research, Wyville Thomson made two discoveries of enormous significance. First, he successfully dredged unquestionable remnants of organic life from a carefully measured depth of more than 600 fathoms—twice the depth at which Forbes's theory had predicted the lifeless zone to start—and second, he found that below about 200 fathoms water temperatures stopped their usual, latitudinally dependent, decrease and took on a life of their own. Wyville Thomson discovered the first firm evidence that the deep ocean was dominated by water currents that have their own temperature and physical characteristics and move under mysterious influences unaffected by conditions at the surface.

So successful was the *Lightning*'s voyage that it was quickly followed by three others: two by HMS *Porcupine* in 1869 and 1870, again led by Wyville Thomson, and one in 1871, conducted by the frigate HMS *Shearwater*. The *Porcupine*'s voyages were notably successful; the *Shearwater*'s was less so, with, as we shall see, far-reaching consequences for the future route of *Challenger*. But even after the *Shearwater* expedition Wyville Thomson and Carpenter had seen enough to be convinced of the scientific effectiveness of deep sea dredging and sounding.

When, in 1870, Wyville Thomson was elected to Forbes's former Chair in the University of Edinburgh, he and Carpenter were influential enough to persuade the council of the Royal Society (then headed by the eminent Thomas Henry Huxley) and the Admiralty to let them organize a much larger expedition. Thus was born the voyage of HMS *Challenger* and an otherwise unremarkable group of naval officers and scientists found themselves walking unawares into scientific history one cold morning in the winter of 1872.

The objectives of the voyage, as finally agreed upon by the circumnavigation committee of the Royal Society were fourfold:

1. To investigate the physical conditions of the deep sea in the

great ocean basins (as far as the neighborhood of the Great Southern Ice Barrier) in regard to depth, temperature, circulation, specific gravity, and penetration of light.

2.   To determine the chemical composition of seawater at various depths from the surface to the bottom, the organic matter in solution and the particles in suspension.

3.   To ascertain the physical and chemical character of deep-sea deposits and the sources of these deposits.

4.   To investigate the distribution of organic life at different depths and on the deep seafloor.

The naval men on board must have been amazed at the speed with which the expedition had been organized. Wyville Thomson and Carpenter applied for funding in 1871, it was approved in April 1872, and by that December the expedition was ready to put to sea! All this because of some success in disproving another professor's pet theory? Of course not. What they did not know was the real reason behind Wyville Thomson's enthusiasm, the Royal Society's rapid deliberations, and the Admiralty's speedy agreement to their request.

Only 14 years earlier the Victorian establishment had been rocked to hear Darwin's ideas about the mutability of animal and plant species by means of his proposed mechanism of natural selection. It was an idea that was still being debated in meeting halls, lecture theatres, salons, and parlors across the empire, but it was gaining wide acceptance among scientists and lay people alike. One of the central tenets—and most significant problems—of Darwin's theory was the burden of proof from the rocks. The fossil record would provide either ultimate affirmation or annihilation of the theory of evolution and the ocean would be its supreme testing ground. Darwin theorized that it was in the ocean that marine organisms found on land only as fossils would still be found alive. Forbes's azoic theory predicted that the very opposite would be found in this abyss of time. It was this investigation that was to form a major part of the *Challenger* expedition. The ship would set out not only to investigate the geology and geography of the sea and seafloor, but also to find the proof of the theory of evolution, or as it was then known, "descent with modification."

It is not fair to say that the voyage of HMS *Challenger* was to be another expedition in the tradition of Darwin's seminal voyage aboard HMS *Beagle* from 1831 to 1836 or Huxley's on HMS *Rattlesnake* from 1846 to 1850, because both of these had been naval voyages with the primary objectives of exploration and territorial annexation. The voyage of *Challenger* was quite different; indeed it was unprecedented in the Victorian world, for it was to be the first voyage sent out with the primary purpose of gathering scientific information. Other nations in the 1860s had begun to appreciate the importance of understanding the sea. The United States, Germany, and Scandinavia had all organized their own expeditions, but notably only Britain had the vision and ability to conceive the idea of funding, outfitting, and deploying such a global mission for purely scientific purposes. Why was this? What made Great Britain so different from the rest of the world? The answer is twofold. First, Britain's economic supremacy was unrivalled in the latter decades of the nineteenth century; it was the largest, wealthiest, and most influential nation on Earth. It could afford to send out a mission for purely scientific purposes and show the world the true meaning of *Pax Britannica*. A scientific naval expedition simply enhanced Britain's prestige in the same way that America's space program would do a century later. For even in the 1860s and 1870s, as sometime Prime Ministers Gladstone and Disraeli fought and bickered in the House of Commons over their humanitarian versus imperial policies, the British Empire's size and hold on the world were still growing; and its sunset was at least four decades away. Second, Britain's economic supremacy was built on a century of maritime pre-eminence. Its merchant marine was the largest in the world as was the navy that looked after it. It simply had the know-how to conduct such an expedition.

But there might have been a third reason, both more fundamental yet less easily articulated. By sending out the *Challenger*, was the Victorian Empire addressing a nagging question that had been spectacularly exposed when "Darwin's Bulldog," Thomas Henry Huxley, clashed with Wilberforce, the Bishop of Oxford, at a meeting of the British Association for the Advancement of Science held in Oxford 12 years earlier? On that occasion, Wilberforce had sneeringly asked Huxley whether it was from his grandfather or his grandmother's side of the family that he claimed descent from an ape. Huxley had easily turned aside the taunt, answering dryly that he would rather have a miserable ape for a grandfather than ridicule reasoned scientific debate. Yet that exchange exposed the schizophrenia of a society that was trying to reconcile God and Science. It was a question that struck right to the heart of that self-satisfied Victorian imperial superiority, the self-belief of a nation that had tamed the world, and which lived and died—at least in public—by a rigid moral code. Was the real brief of the *Challenger* expedition nothing less than a last chance to choose between God and Science? If it was, perhaps that explains why to the Victorians the *Challenger* expedition was every bit as important as the *Apollo* moon landings would be to another great nation a century later.

TO BOLDLY GO . . .

Sheerness, England, November 22, 1872, 51 ° 27′ N, 00 ° 45′ E

HMS *Challenger* was built in the Royal Naval Shipyards at Woolwich and launched on February 13, 1858, the same year that the first transatlantic telegraph cable was completed. Indeed the advent of the telegraph was to be hugely important for the success of the *Challenger* expedition because, via this "Victorian internet," the expedition's leaders provided regular progress reports to satisfy the voracious appetite for science back home. *Challenger* was a spar-decked, three-masted corvette with a modestly powered auxiliary steam engine that engaged with a twin-bladed propeller assembly,

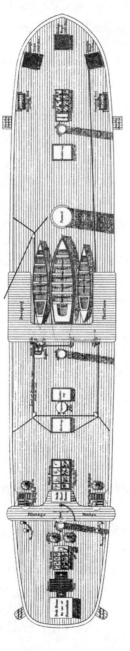

H.M.S. CHALLENGER,

as fitted for a voyage of Deep sea exploration. Dec.r 1872.

UPPER DECK.

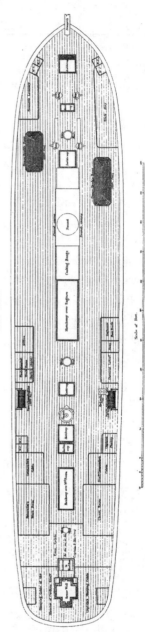

MAIN DECK.

Scale of Feet.

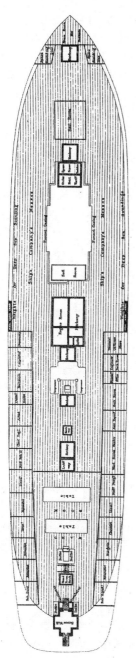

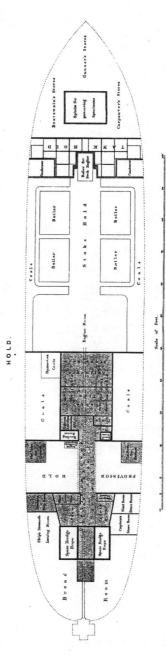

FIGURE 1 Plan of *Challenger*

which could be disconnected and hoisted clear of the water when the ship was under sail. Technologically she was a hybrid that straddled the eras of sail and steam. In 1861 she left for her first foreign tour on the east coast of North America and in the West Indies, taking part, in 1862, in operations against Mexico. Her next commission was in the South Seas on the Australian station (as it was called) where she visited Fiji on a punitive mission to avenge the death of a missionary and his dependents who had been murdered by natives. This was typical of Victorian foreign policy, which ruled the empire by naval action, though often just its threat was sufficient.

Even by the standards of the day, *Challenger* was not large, displacing only 2,300 tons and being only 200 feet long. But this was, as Wyville Thomson commented, an advantage, because she had all the extra space and amenities of a frigate combined with the maneuverability and draught of a corvette. All but two of her original 17 guns were removed to make room for laboratories and storage cupboards, the cabins for the scientists, and the miles and miles of hemp dredging rope and steel piano wire that were to be used for sounding. The funnel dominated the ship. Fully 10 feet in circumference, it was the exhaust for the relatively small and very inefficient 1,234-horsepower steam engine. Amidships on the upper deck was the dredging platform itself, flanked by zinc specimen boxes and with a small steam donkey engine to one side to pull up the dredge with its precious cargo of samples. Figure 1 shows the layout of the ship.

This engine drove an axle that ran clear across the ship. For'ard were berths for three small boats, the gigs for additional sampling (see Figure 2), going ashore, or rendezvousing with other vessels at sea. Below was the cramped main deck with an enormous cooking range that dominated the center of the ship. Around an open central area were arranged the cabins for the senior officers and the scientific team with the captain's and principal scientist's berths situated aft near a main laboratory that was dedicated to drawing, describ-

FIGURE 2 Instrument and sampling platform

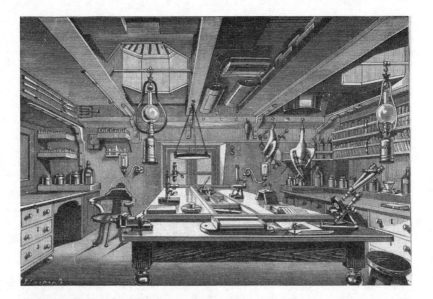

FIGURE 3 Naturalist's lab

ing, cataloging, and preserving specimens. Figure 3 shows the
naturalist's lab aboard *Challenger*. This main deck was dimly illumi-
nated, lit only by three small skylights during the day or by oil
lamps at night. Below the main deck was the lower deck with cabins for the junior officers and the berths and messes for the ship's
company—the bluejackets and marines—even darker and more
poorly ventilated. With a total crew complement of 269, *Challenger*
was very cramped—even the junior officers had to share cabins—
and it is not surprising that in the course of that four-year voyage
fully a quarter of the seamen aboard deserted, especially when
tempted by such exotic locales as South Africa and Australia. Below
the lower deck was the hold, with storage room for food and coal,
additional dredging and sounding rope, and the engine and its four
boilers.

 Challenger was commanded by Captain George S. Nares, one of
the greatest surveyors in the navy, who in later years became famous
as an Arctic explorer. He commanded 23 naval officers and a crew

of 240 ratings and able seamen. Wyville Thomson was the chief scientist, assisted by a scientific staff of five. One of these was John Murray, a fiery and outspoken Canadian who was to become the most famous of all *Challenger* scientists as the lead author of the massive 50-volume tome that eventually "summarized" their findings. Murray had an overwhelming interest in natural history and was prepared to rough it for weeks or months on end—on land or at sea—in pursuit of interesting specimens. However, his interest in natural history was not matched by a parallel interest in conventional scholarship. As a student, he commonly missed lectures and never attended examinations, preferring instead to work hard at any subject to which his eclectic interests led him. He was that rarest of scientists, a synthesist capable of seeing what we now call "the big picture." Murray, too, was an Edinburgh product and it says much about that University's far-sighted system that it let him have his own way. "Having his own way," in fact, was something at which John Murray excelled, for as well as being a talented intellectual he was also a brilliant entrepreneur and in later years amassed a significant fortune by combining his talent for science with an innate business acumen. He exploited the Christmas Island guano deposits for fertilizer. But at the start of *Challenger's* voyage Murray, only 31 years old, was merely another of the young scientists attracted by Wyville Thomson's brilliance and reputation.

Another such scientist on board *Challenger* was Henry Nottidge Moseley who, like most of the other scientists and crew, was a young man in his 20s when the expedition departed. He was born in Wandsworth, London in 1844 and early on developed a love of natural history that was to stay with him for the rest of his life. Although Moseley was to become one of the greatest natural scientists of his generation he was not, as were many Victorian scientists like Darwin and Huxley, wedded to science to the exclusion of all else. His interest in travel and natural history seems to have owed as much to an early love of Defoe's *Robinson Crusoe* as to any formal education. He was educated at Harrow school where he did not stand out as either scholar or sportsman. Yet he loved to fool around in his small homemade laboratory and was able to produce smells there of such extraordinary potency that his normally indulgent headmaster feared he would have to put a stop to them on the grounds of sanitation. In 1864 Moseley went up to Exeter College, Oxford to take a degree that was to be in either math or classics (his father was both a canon of the Church and a noted mathematician). But Moseley was enormously unhappy. He was cut out neither for religion nor for the dry minutiae of algebra and he idled away his days in long country rambles around Oxford, collecting curiosities for his own natural history collection. It was in this state that he was discovered one day by an old—and rather more liberal—friend of his father's. This friend, also a clergyman, realizing that Moseley was wasting his life in Oxford, interceded on his behalf with George Rolleston, professor of anatomy. Moseley was enrolled in the recently established Honor School of Natural Science at Oxford, where he immediately blossomed. He won a first-class degree in natural sciences in 1868 and, after a four-year dalliance with a career in medicine, was chosen for the *Challenger* expedition.

It was a moment of high excitement for Moseley who, as a boy, had been much influenced by Darwin's book *The Voyage of the Beagle*. He spent several months fitting out a state-of-the-art zoological laboratory on board the ship before boarding it in preparation for departure at Sheerness in November 1872. He was a short rather stoutish young man with a luxuriant black moustache that hung down to his chin. He was also, as his companions were soon to discover, immensely kind and sympathetic and had enormous energy and enthusiasm for the voyage ahead.

There was one lonely young man in particular who was pleased to meet the likeable Moseley. At 25, Rudolf von Willemoes Suhm was the youngest of the "Scientifics" (as they came to be dubbed by the crew) to be recruited to the voyage. He had started his intellectual life thinking that he was destined to be a lawyer but had spent so much of his first year at the University of Bonn indulging his interest in natural history that he soon changed to major in zoology. He took a precocious doctor's degree only two-and-a-half years later, by which time he had developed a deep interest in marine invertebrates. His career was interrupted by military service but then he returned to science, this time to the University of Munich as an assistant in the zoological museum.

His involvement with the *Challenger* expedition began quite by accident, and very late on in the preparations. He met Wyville Thomson in October 1872, when the German survey vessel *Phoenix* put into Edinburgh to coal. He and Wyville Thomson hit it off so famously that Wyville Thomson asked him there and then if he would care to join the *Challenger* expedition. The young von Willemoes Suhm was so overwhelmed by the invitation that he immediately took a leave of absence from the *Phoenix*. Five days later, and on Wyville Thomson's instruction, von Willemoes Suhm traveled by land to London for an appointment with the doyen of Victorian science, Thomas Henry Huxley, in his lair in the South Kensington Museum. Huxley greeted him with great warmth and promised to use his influence with the circumnavigation committee to secure von Willemoes Suhm a place on the expedition. Less than a week later, having rejoined the *Phoenix*, von Willemoes Suhm received a telegram in Copenhagen from the Admiralty informing him that he had been appointed a naturalist on the expedition. It was the crowning moment of the young German's life, an entrée into a glittering career in science of which his scientific peers could only dream. With no inkling of the tragedy that awaited him he gleefully accepted this once in a lifetime invitation.

At 29 John Young Buchanan was slightly older than von Willemoes Suhm when *Challenger* departed Portsmouth. Like von Willemoes Suhm, he had come to science relatively late, having started a degree in Glasgow to read arts before discovering a deep-seated love of chemistry. After studying on the Continent he returned to Scotland to work with Crum Brown, the noted professor of chemistry at Edinburgh. Buchanan was a superbly practical chemist, capable of making his own instruments and skilled in the art of glass blowing, something that was likely to be in great demand on a naval vessel braving some of the roughest waters in the world. Buchanan was a kindly man, very sincere, but so shy that his sunny nature was almost never apparent. However, when he did form friendships they were deep and long lasting and this was the case when he met the young German, von Willemoes Suhm, who was feeling very alone and very young when he, too, joined *Challenger* at the end of November 1872. Like von Willemoes Suhm's other new friend, Henry Moseley, Buchanan established a well-equipped laboratory on board, for despite the overarching imperative to find the proof of Darwin's theory of descent with modification, the physical sciences were no less regarded than the natural sciences on the expedition. Consequently, one of the gun bays on the main deck was converted into a tiny but serviceable physical and chemical laboratory for Buchanan's use.

The final member of the Scientifics who joined *Challenger* at Sheerness—John James Wild—is today something of an enigma. He was Wyville Thomson's secretary and was allocated part of the great man's cabin in the aft of the ship. Wyville Thomson wrote "[The port-end of the fore-cabin] being appropriated to my use and that of my secretary, Mr. Wild, to whose facile pencil we are indebted for beautiful illustrations of our novelties, and who sits with me in gathering the various threads which we combine into a symmetrical web as best we may."

A month before *Challenger* arrived in Portsmouth to be readied for its departure on its long voyage, it finished a complete refit in the naval dockyard at Sheerness. It was here that all but two of its 17 68-pound guns were removed and the rest of the conversions— such as Buchanan's and Moseley's laboratories—were completed, thus changing an unremarkable corvette into the first dedicated scientific exploration vessel in the history of the world.

But important as the laboratory fitments and the scientific staff were to the future success of the voyage there was another, larger,

group on board that was arguably even more crucial. These were
the officers and crew (the latter known as "tars" or "bluejackets")
who would actually look after the command, navigational, and
housekeeping chores without which the scientific work would be
impossible. Three of the officers published accounts of their voyage
on their return, and one of these even made it onto the late Victorian bestseller lists. From the perspective of the Victorian public,
perhaps the most notable of these literary officers was the aristocratic Lord George Campbell, youngest son of the eighth duke of
Argyll and a sub-lieutenant on board *Challenger*. He was tall and
lean with a well-trimmed dark beard and a sardonic sense of humor.
In physical appearance he was a perfect counterpoint to Navigating
Sub-lieutenant Herbert Swire. Swire was unbearded, blond, and
some years younger than Campbell but, like Campbell, had an
irreverent sense of humor. It was Swire who, in his diary, unpub-
lished until after his death, named *Challenger*'s earnest scientific staff
the "Philosophers." Swire was particularly amused by the clothing
that the Philosophers chose to wear: formal gentlemen's garments
with waistcoats and watch fobs that would have looked at home in
Pall Mall but that were anything but suitable for the deck of a small
corvette. But he was an intelligent young man, hard-working, with
a keen eye and an appreciation of the finer things in life, and it was
he who wrote most lyrically about the colors, sunsets, and lands
that they saw on their long voyage. More practical, but with a flair
for detail, were the writings of Engineering Sub-lieutenant William
Spry, a man who spent much of his time tending *Challenger*'s tem-
peramental steam engine. It was Spry's book that became the best-
selling account of the voyage when he returned to England, running
into more than 10 editions by the end of the nineteenth century.

Until very recently we had no record of life below decks on
Challenger. Indeed there was no expectation of one, because the
average tar was hardly noted for his literary ability. Many of those
who shipped below decks in the service of the empire could not
even read, despite the educational reforms that were even then

beginning to sweep Britain. But incredibly, on board *Challenger* there was one seaman who left an account of the voyage in 69 letters to his family and friends at home. Until 1985 there was no hint that these letters existed, but in that year his granddaughter gave them to the Scripps Institution of Oceanography in La Jolla, California.

Joseph Matkin, a short dark-complexioned lad with dark hair and pale blue eyes that missed nothing on their long trip around the world, was the ship's steward's assistant. He was only 18 years old but had received an excellent education at a good school near his home in Oakham (a town in the small county of Rutland). The quality of his education there was vital for his future literary efforts; so was his upbringing by parents who wanted their offspring to enjoy all the benefits of the education that they had not had. So young Joe Matkin was very much a child of his times, plugged into the new Victorian ethos of self-betterment.

By the time he was 12, Matkin had left school and enrolled, somewhat surprisingly, in the merchant marine. In that service he sailed for Australia aboard *Sussex*, returning the following summer aboard *Agamemnon*. Not long after that he sailed for Australia again, this time aboard *Essex*, remaining in Melbourne for a year. He was back in England by 1870 and decided to enlist in the Royal Navy, where he served as ship's steward's boy aboard HMS *Invincible* and HMS *Audacious*. Life in the Royal Navy had improved quite a bit since the privations suffered by seamen in the 1850s during the Crimean War—with less risk of being "pressed" and better controls on the use of punishments employed in the old canvas-and-tar navy to maintain discipline. However, life afloat in the Victorian sail-to-steam transitional navy was still not notably comfortable and *Challenger*, with its several laboratories and huge storerooms, was even less so than Matkin's previous vessels. However, he was a bright and ambitious boy and it was the promise of advancement—as well as an all-expenses-paid trip around the world—that brought him on board three weeks before his nineteenth birthday on November 12, 1872.

Challenger put out from Sheerness on December 7, 1872 in the shadow of recent tragedy. At seven o'clock in the evening of Monday, November 18, a young marine, Tom Tubbs, was negotiating the ship's gangway in the treacherous darkness of the unlit dock-yard when he missed his footing. It was a straight drop into the rank filthy waters of the Dock Basin and he sank without trace in his heavy clothes, drowning in front of his shipmates before they could find a light or throw him a line. All night his mates and the dock police frantically dredged the area where he fell under the glare of hastily rigged lights. But their efforts were to no avail and at first light a diver was called. With nothing but the clank of the hand-operated air pump to disturb the silence, everyone on board *Challenger* looked on, as the diver, in his iron-weighted suit, was lowered over the side into the scummy water. For several minutes he searched without success until suddenly he came upon Tubbs' body propped bolt upright against the side of the 27-foot-deep basin, his sightless eyes staring off into the blackness.

So it was a subdued company, muttering nervously about ill omens, that set out from Sheerness a few weeks later, en route for Portsmouth, where they would load the last of the stores. Their mood—and their misgivings—were not improved by the ferocious southwesterly they encountered as soon as they left the harbor. So violently was the ship tossed around that only two days after leaving Sheerness *Challenger* had to put into Deal in Kent, where—to a man—the scientific staff suddenly rediscovered the delights of rail travel. They arrived in Portsmouth two days before the ship that was to be their scientific and spiritual home for the next three-and-a-half years.

It was an inauspicious message to send to their shipmates and was received with appropriate contempt by the officers and bluejackets who had sailed the ship on to Pompey (the colloquial name for Portsmouth in the British navy) alone and had been forced to put in twice more because of foul weather. *Challenger* finally arrived at Portsmouth on Thursday, December 12, 1872. In the time it took the ship to travel those trivial 105 miles *Challenger* made less than 1 knot against the headwind—and that was under the full power of her steam engine. That was not the only abuse that the ship's primitive engine endured so soon after leaving home; at the height of the storm, the heavy sea found its way through several unsecured hatchways into the engine room. If not for the fast work of Engineer Spry and his colleagues, the boilers themselves would have been doused with water, risking an explosion that could have destroyed the ship.

On Saturday, December 21, 1872, *Challenger* finally left Portsmouth, having taken on so many additional supplies that the holds seemed, as William Spry put it, "scientifically constructed of some elastic material so as to stretch to any size." As soon as they reached the chops of the Channel they were once again in heavy weather and *Challenger* and her crew, who had not yet had the time to find their sea legs, were pitched around by the swell. Weighing heavily upon them too was the knowledge that they would not see the shores of Britain again for four long years.

In his cabin, Wyville Thomson reflected sadly on the wife and son he was leaving behind and wondered again whether he had been right to accept the circumnavigation committee's invitation to lead the expedition in the face of William Carpenter's obvious wish to do so. In fact, the last few days at Portsmouth had been made very difficult by Carpenter's blatant expression of anger and humiliation at not being selected. It was a sad end to the fertile relationship that had led them from a common love of the lower animals in university zoology labs to the commissioning of the *Lightning, Porcupine,* and *Shearwater* expeditions and then the organization of this, the ultimate voyage of scientific discovery, the *Challenger* expedition itself.

Down in the engine room, as the pistons pounded and the regulators spun into a blur, William Spry wrote of the "cherished recollections" that would need to sustain him in the long task ahead and tried to contain the "melancholy impressions" that besieged him. In the

Chief Petty Officer's mess, Joe Matkin had problems beyond his own sense of loss at leaving England. He and the six companions who shared the mess with him were obliged, as all ratings were, to kit their mess out with cutlery and crockery at their own expense. After that and the purchase of some necessary items of clothing, his advance looked very paltry. Now he would not be able to pay back the 10 pounds his father had lent him until he reached New York in the early summer of the following year. For an honest lad who knew that his father's health was failing, the omission was all but unbearable.

Despite all the hardship and bad luck, on December 25 the crew tried to make the best of their first Christmas aboard. The Captain led the worship, for ships with fewer than 295 hands did not rate their own chaplain. Christmas dinner in all the messes was a miserable affair. In the midst of a storm with a heavy swell, the diners found themselves constantly having to hang onto their crockery while those with weaker constitutions were confined by seasickness to their cabins. One wave was so heavy that John Buchanan was flung from his seat out of the officer's mess and into an adjoining cabin! Yet it was the officers who came off worst that first Christmas Day, for just before dinner, at a few minutes to six o'clock, their turkey mysteriously disappeared from the galley. The next night, on Boxing Day, their roast goose was similarly appropriated and all that was found of it were some scraps of meat and flakes of salt up in the main top rigging. So as *Challenger* crossed the Bay of Biscay—not far from the spot where the navy's new ironclad HMS *Captain* had gone down with all hands only two years before—her crew battled not only with the weather, their homesickness, and their cramped conditions, but the knowledge that somehow—along with all their extra provisions—they had shipped a thief, too.

THE FLOATING WORLD

As *Challenger* headed south for Lisbon and Gibraltar, it crossed the European continental shelf. It was an area that had already been

studied by the *Porcupine, Lightning,* and *Shearwater* and so was home
ground to the Scientifics aboard *Challenger.* They looked forward to
leaving this familiar territory and entering the realm of the truly
unknown beyond Gibraltar. But continental shelves have their own
fascination. They make up more than 8 percent of the world's total
oceanic area and the European continental shelf is one of the largest
in the world, swelling outward from the Spanish port of Santander
in the southern part of the Bay of Biscay and extending far north to
the edge of the deep Arctic Ocean.

Technically, continental shelves are defined as the region of
ocean floor between the coast and the shelf-break, where the seafloor steepens into the continental slope and plunges toward the
abyssal depths. Most shelves have a gently rolling "ridge and swale"
topography and slope gradually toward the shelf-break at inclina-
tions of not much more than a tenth of a degree per kilometer.
Because they are so shallow—typically less than 300 meters deep—
continental shelves are very susceptible to the variations in sea level
caused by the waxing and waning of the ice sheets. Indeed the
noted American oceanographer, Don Swift, has said that continental
shelves can be thought of as ancient parchments continually written
on by geology and yet erased by the movement of the ice sheets
and sea level. Technically, continental shelves are continuations of
the continents, because they have the same geology—the outcropping (that is, surface) rock beneath the sea is the same as the
outcropping rock on the adjacent land—and because they are made
up of a type of crust called continental crust. It has only been in the
last 40 years that the difference between continental and oceanic
crust has come to be appreciated as one of the most fundamental
geological distinctions on our planet.

To understand just how important continents—and therefore
continental shelves—are, we need to know a little planetary geology.
There are three types of planetary crust. "Primary crust" is planetary crust that was formed at the time the solar system was formed.
It is common on dead planetary bodies—a good example being the

light-colored highlands of our Moon. "Secondary crust" is that formed by the action of heat generated by the decay of radioactive minerals inside planets. This heat gradually accumulates and eventually causes localized eruptions of basaltic magmas (molten igneous rock). The surfaces of planets such as Mars and Venus are of this type and, crucially, the ocean floor of our own planet is also constructed in this way. The third type, "Tertiary crust," is, as far as we know, unique to our home world. It forms when surface rock is continuously recycled back into the interior of a planet by the processes associated with geological activity.

"Geologically active" in Earth science circles implies the involvement of the phenomenon known as plate tectonics, the processes by which new rock is formed at the mid-ocean ridges and subducted at the ocean margins. We will see just how importantly the *Challenger* expedition contributed to the understanding of plate tectonics. For the moment, though, it is enough to think of this continuous production and consumption of crust as a form of distillation—not in this case of whisky or gin—but of rock itself. The normal type of igneous or "volcanic" rock on our planet is basalt, a hard, black, featureless, and heavy rock that underpins the sediments of the oceans. But continental crust—formed by this endless process of distillation—is quite different; it is a paler gray in color, lighter in weight, and thicker. This means it floats higher than oceanic crust and continents stand proud of the oceans. All crust floats on top of the eternally roiling cauldron of the fluid mantle, but oceanic crust floats lower. Think of two bottles of shampoo floating in the suds during the kid's bath time. One is full and so floats lower in the water than the one that is half full—that's the difference between oceanic and continental crust.

What, then, is the singular process that produced this strange duality between oceanic and continental crust—a difference that, as far as we know, is not duplicated on any other planet in the solar system? A crucial factor is the *rate* of the distillation process—the rate of cooling of the magma. On volcanically active planets like

Venus, the surface roils and churns so quickly that solid crust does not have a chance to form. On geologically static planets like the Moon, geology was a game that was played only once, very quickly after the planet was formed, and then was gone forever. NASA's *Apollo* missions showed very clearly that the only geological action now on the moon comes from the eternal impact of dust particles captured by its indifferent gravity as well as the occasional apocalyptic arrival of an impacting asteroid or planetoid.

Only Earth, as far as we know, has the compromise necessary to produce continents—the internal fires burn quickly enough to bring magma to the surface and, as it spreads away from the mid-ocean ridges, it has the necessary time to cool. It is then recycled back into the interior of the planet at the ocean margins where, like a sinking slab of toffee, it is returned to a molten state. This mass of molten continent still, however, retains enough homogeneity to be returned more or less as a unit at the mid-ocean ridges eons later as the rock cycle continues. Then it starts the same process all over again. With each iteration—each distillation—the chemical composition of the heated magma changes, becoming lighter and lighter, so that in time these great slabs of rock ride higher in the magma ocean than the oceanic crust. The continents are like the ships of the Victorian navy, patiently constructed to ride the oceans of the inner planet.

Looked at this way, with the advantage of a century and a half of geological research that *Challenger's* Scientifics could not have imagined, we can see that as our iron-and-wood corvette made her way across the Channel and out into the storm-racked Bay of Biscay she had not even started her long voyage of discovery. Her officers, bluejackets, and Scientifics were finally at sea, yet from the perspective of geology they had not yet even left the continents. The true silent landscape still lay before them.

The Desert Under the Sea

Lisbon, Portugal, January 3, 1873, 38° 44′ N, 09° 08′ W to
Gibraltar, Mediterranean Sea, January 18, 1873, 36° 09′ N, 05° 21′ W

THE LOST EMPIRE

On January 3, 1873 *Challenger* entered Lisbon Roads on the river
Tagus, arriving at the city of Lisbon at midday. On either side of her
rose hills covered with vineyards and the gently rotating white sails
of the windmills used for crushing grapes. Before she was allowed
to anchor, the Portuguese admiral's boat came alongside and inter-
rogated Captain Nares closely: "The name of this ship, sir? It's purpose, your armament, hands on board, any sickness?" Despite their
exhaustion, the *Challenger's* crew were both amused and infuriated
by what they saw as continental officiousness. Who did they think
they were, addressing a vessel of Queen Victoria's empire that way?

The men of the *Challenger* expedition were a product of an
"empire on which the sun would never set." In word, thought, and
deed they were colonialists steeped in unconscious prejudice. Young
Joe Matkin observed that the Portuguese seamen looked dirty and
that even their flagship was not as good as his own modest *Challenger*. William Spry observed smugly that although the churches,
gardens, and palaces scattered about were worth a visit, the full tide
of Portugal's prosperity had receded with their empire 300 years
earlier. Yes, the Portuguese had been great explorers; after all was it
not Vasco de Gama who sailed to the Cape of Good Hope and then

visited India? Other Portuguese explorers had conquered the Maldives, established industries in Ceylon, the Moluccas, Sumatra, and other places on "the Eastern Archipelago." The Portuguese had established early trade links with China and Japan, connections that the British Empire had been able to forge only recently. Indeed, in the early sixteenth century no flag but that of Portugal flew along the whole length of the African coast and no ship dared anchor in any harbor from Gibraltar to Abyssinia, Ormuz to Siam, without the permission of the Portuguese. In the sixteenth century, England could not have disputed the possession of an inch of ground with Portugal for a week, but who ruled India now?

The rot, according to Spry, set in 1557 when misgovernment, tyranny, the Jesuits, and the Inquisition strangled the rising fortunes of the Portuguese. Now Portugal was stripped of nearly all its colonies, a shadow of what she had been. "England," concluded Spry, "now wears the mantle Portugal in her blindness and bigotry let fall."

Yet after a few days in Lisbon, the men of *Challenger* found the place growing on them. Despite the pervasive smell of garlic on the local breath, which particularly disgusted Matkin, they had to allow that fruit and fish were fresh, plentiful, and cheap and made a welcome change from the monotony of shipboard rations. Culturally, too, there was much to admire in Lisbon. Herbert Swire enjoyed a visit to the opera while Matkin and a mate told of bullfights, theatre, and a masquerade ball where even the lowliest rating could dance with a countess. And could it be that the Portuguese were not so ignorant after all? Just before departure, the King of Portugal, an enthusiastic natural historian, paid a visit to the ship, where Nares and Wyville Thomson, impressed with the king's knowledge of biology, took great delight in introducing him to *Challenger's* laboratories.

After being detained in Lisbon for two days longer than planned because of bad weather, on January 12, *Challenger* made all plain sail for the most southerly of Victoria's possessions in Europe, Gibraltar.

NIGHT OF THE LIVING DEAD

Challenger's Scientifics might have had no inkling of the true nature of continental shelves yet they did make some significant discoveries on this leg of the voyage. Soon after leaving Lisbon, in water 1,000 fathoms (6,000 feet, a little more than a mile) deep, they dredged a sea lily. Technically, sea lilies are known as crinoids and are a member of the same phylum as starfish and sea urchins although they look very different, having many articulated arms connected to a long stalk that anchors the animal to the seabed. Arms and stalk are covered with hard plates of lime (calcium carbonate). The crinoids have a long and illustrious fossil record and were exceptionally abundant in the Jurassic period at the height of the dinosaur's reign.

The Scientifics' recovery of the crinoid was significant because it proved that the *Challenger* expedition was fulfilling one of its primary roles: testing Darwin's theory that the bottom of the ocean was a haven for life forms found on land only as fossils. Darwin's theory of "descent with modification" stated that the variation among offspring (produced through genetic recombination, a mechanism at that time not fully understood) was then worked on by the process of natural selection. "Natural selection" was a term that Darwin coined to indicate the winnowing of unsuccessful body types and chracteristics through early mortality while better-adapted types developed into successful species. This "differential reproductive success" is another way of saying that better-adapted organisms leave more offspring behind. These offspring, too, leave more offspring behind and so on. The process continues *ad infinitum* as the successful species out-competes less well-adapted forms. This process fuels an eternal struggle to optimize an organism to its environment, with the implication that a successful species will eventually be perfectly adapted to its environment. In the language of ecology, it will have successfully and completely occupied an ecological niche.

Because the environment itself has continually changed over geological time, the floras and faunas of Earth have had to continually change too. Successful adaptation is like trying to score a goal while the goalposts are constantly moved. But the Victorians believed that on the unchanging ocean floor—the silent land-scape—organisms would not have been forced to evolve beyond the form best fitted to that environment millennia ago and, there-fore, that the animals of the deep would be evolutionary throw-backs, living fossils. This idea also highlighted another of the Scientifics' preconceptions: namely that the ocean floor is the most ancient place on Earth. Both of these ideas would eventually be shown to be wrong, but it would be the *Challenger* expedition itself that would overturn the "evolutionary throwback" notion.

But when they dredged up the sea lily, the *Challenger* crew were simply elated to have found what they thought was the proof of Mr. Darwin's theory, that the ocean floor was indeed a dark, ancient, and unchanging place, populated by the living dead. To the crew, this success seemed like a long overdue good omen for the rest of the voyage.

THE ECHO OF AN IDEA

Sounding and dredging were the two techniques at the heart of *Challenger's* enterprise; sounding to measure the depth of the ocean over which the ship passed and dredging to bring up material for study. Today these seem incredibly primitive, just throwing a string over the side of a boat. But in fact they represented the cutting edge of Victorian remote sensing technology. The sounding principle had reached its highest form of development in 1853, just 19 years before *Challenger* sailed, through the efforts of one John Mercer Brooke. Brooke had developed a method that solved the two main problems of sounding with a line: He used twine rather than hemp rope, so that its weight would not continue to pull more of the line overboard after it hit bottom; and he used a weight to keep the twine as vertical as possible so that inaccuracies due to the curva-ture of the line as the ship moved would be all but eliminated. In

fact the weight was quite sophisticated. It was an iron ball connected to a rod-like apparatus that allowed the ball to be detached
and the twine pulled up without snapping. A further innovation
was a device on the rod assembly that took a sample to prove that
the sounder had hit bottom. This system was later used successfully
by the U.S. Navy to take soundings right across the Atlantic Ocean.

Despite the fact that occasional soundings in the vastness of the
Atlantic were totally inadequate to generate a true picture of the
subsurface topography, in 1854 the director of the U.S. Navy Depot
of Charts and Instruments, Matthew Maury, published the first
"chart" of the undersea geography of the Atlantic. It successfully
showed the steepening where the continental slope began and also
hinted at the presence of a plateau in the middle of the Atlantic—a
feature that was soon to assume overwhelming importance to science. This feature Maury named the "Dolphin Rise" after the ship
that had carried Brooke's cutting-edge sounding gear. But physical
sounding with a line never had much of a chance as a means of
mapping the silent landscape—better eyes were needed. Strangely
enough, when they finally arrived, in the early twentieth century
after *Titanic's* epochal encounter with an iceberg, it was through a
technique, echo sounding, which had been developed at the start of
the same era that had spawned the *Challenger* expedition.

In 1838 Charles Bonneycastle, a young professor at the
University of Virginia, was interested in an experiment with sound
and water that had been performed 12 years before by the Swiss
mathematician Jean Daniel Colladon. Colladon and an assistant had
measured the time needed for sound to travel through water by
striking a submerged bell and releasing a flare at the same time.
From this, Colladon calculated that sound traveled underwater at
almost a mile per second—four times faster than the speed of sound
in air. It was this idea that Bonneycastle used 12 years later when he
tried to estimate the depth of the ocean floor using sound. A team
of sailors from the U.S. Navy brig *Washington* detonated a mine
while a little way off Bonneycastle sat with the narrow end of a

flared tin pipe in his ear. He heard the sound of the detonation and a fraction of a second later another, smaller explosion which he took to be the echo of the main blast off the seafloor. Using Colladon's estimate of the speed of sound in seawater, Bonneycastle was able to calculate the depth of the ocean at that place: about 160 fathoms (960 feet). But when he checked his estimate using a sounding line he found that the actual depth there was only about 540 feet. Discouraged, he abandoned his attempts and for the rest of the nineteenth century the echo-sounding technique was forgotten.

It took the tragedies of *Titanic* and *Lusitania* in the early twentieth century to renew interest in echo sounding as a way to detect unseen obstacles at sea. By then the Submarine Signal Company of Boston, Massachusetts had for some years been manufacturing underwater bells that could be attached to ships and microphones that picked up their peals. In this way two ships that were out of visual range could detect each other's presence.

One of the company's more innovative employees, Reginald A. Fessenden, took this approach a step further and developed a bell so loud that its *reflected* echo from an underwater hazard was audible without amplification. The bell was actually a large metal membrane warped by the action of electrical current. It was so effective that on field trials off the Grand Banks of Newfoundland officers eating in the ship's mess could hear the echo of the device reflected from an iceberg two miles away and, crucially, also from the ocean floor, one mile straight down. The Fessenden Oscillator became the sound source used by U.S. Navy physicist Harvey C. Hayes to develop, a few years later, the first true echo sounder. It was Hayes who overcame the final problem of echo sounding: measuring the time interval between the transmitted and received sound pulses with sufficient accuracy. This measurement was crucial, because a time error of only half-a-second translated into a depth error of more than a thousand feet. His technique was to "center" the sound, that is, to vary the outgoing pulse frequency until it coincided exactly with the incoming echo. To accomplish this the operator used a set of headphones, one earpiece registering the outgoing pulse, the other, the incoming echo. When the two noises coincided exactly the distance to the seafloor was simply the time difference multiplied by the speed of sound in seawater, 0.9 miles per second. To find the time difference the operator detuned the device by varying the frequency control—speeding up or slowing down the succession of pulses—until once again the two sound pulses coincided. The time-distance equation could now be solved using basic algebra and the echo sounder—or depth fathometer—was born.

It was now possible to make almost continuous measurements of the ocean depth along a ship's track, and Hayes himself was the first to do this in 1922 while sailing between Newport, Virginia and Gibraltar aboard the USS *Stewart*. During the single week that he was at sea he managed to take more than 900 soundings—more than *Challenger* had managed in its entire four-year voyage. From then on, the development of the fathometer was continuous, initially embracing automated pen recorders to graph changing submarine topography and then being integrated with solid-state electronics and computers. This integration ultimately gave rise to a technique that came to dominate the world of underwater warfare—sonar (the term stands for SOound NAvigation and Ranging—but it is certain that this acronym was created retrospectively. It is likely that the original term was coined simply because to early operators it was perceived to be "something like radar").

The impetus for developing sonar was the limitation that the Hayes Fathometer (and its other variants developed in the years after the First World War) could reconstruct only the topography of the seabed directly underneath a ship's track. The fathometer's output is a one-dimensional line and the ocean floor is a two-dimensional surface. An additional problem was that the width of the line was not clearly defined—the sonar "ping" spreads on its way down to the ocean floor as well as on its way back up, resulting in a depth estimate that is an amalgam of the topography beneath the ship.

To get around these difficulties, Harold Farr and Paul Froelich, of the Harris Anti-Submarine Warfare division of the General Instruments Corporation of Massachusetts, developed multibeam sonar. Their technique depends on the use of interference patterns, the physical phenomenon whereby light or sound waves, when passed through a diffraction grating, either disappear by canceling each other out or augment each other. Stringing a row of several sonar transmitters beneath a ship in an array from bow to stern makes the sound waves fan out, mapping the seafloor beneath and on either side, but from back to front they cancel each other out. This "insonification" phenomenon results in a stripe of sonar pings coming back to the receivers for miles to port and starboard but not from fore and aft.

The real genius of Farr's and Froelich's invention, though, was to string the array of *receivers* at right angles to the transmitter array. A signal originating from directly beneath the ship would reach all the receivers at once but a signal originating from, say, far to port would reach the port side of the receiver array first and then progressively along each receiver until it reached the starboard-most one. In this way, Farr and Froelich were able to calculate not only the delay between an outgoing signal and its arrival back at the receiver, but the tiny time differences between its reception along the elements of the array. From these results they would then calculate not only the distance to the echo source but also its angle underwater. At last a tool had been devised that could accurately measure the depth of features on the seafloor on either side of a ship as it moved forward. With electronic integration, this swathe or line on either side could be easily turned into a map of the surface of the seafloor.

Multibeam sonar was given to the U.S. military early in the 1960s and they immediately did what the military always do with an innovation—they classified it Top Secret. But after 30 years even the U.S. Navy could see that, following the collapse of the Cold War, there was not much point sitting on an invention that would finally allow us to visualize the 70 percent of our planet that remained almost completely unknown and so they allowed a commercial version called Sea-Beam sonar to be developed. But this instrument is a pale reflection of the full-blooded military version, having only about a third as many sensors as multibeam sonar. Nevertheless, Sea-Beam sonar, combined with accurate satellite navigation, finally allows us to map the topography of the silent landscape and it is this development, as we shall see, that has confirmed one of the most extraordinary discoveries of *Challenger's* voyage, the mid-ocean ridges.

But multibeam sonar was as remote and incomprehensible to the crew of HMS *Challenger* as the dark side of the moon. They had to be content with their sounding lines of piano wire. Their methods of physically retrieving samples from the ocean floor were similarly primitive. Their dredge was an iron frame that held open the mouth of a bag made of finely woven material. It was connected to the ship by one-inch-diameter hemp ropes with a drag so fierce that it typically took more than three hours to lower the dredge to a depth of 2,000 fathoms. Once the dredge was on the bottom, the ship stayed on station all day, hauling the dredge slowly along under steam power. To prevent the dredging line from snapping, a special "accumulator" made from gutta percha (india rubber) ropes connected the line to the supporting mast. The accumulator absorbed severe variations in tension and protected the mast. Transits between dredging sites had to be done under sail in order to conserve fuel. After the day's dredging, the haul was lifted from the depths by the ship's tiny 12-horsepower donkey engine; then the eager scientists fell upon the day's pickings and worked far into the night in their odorous laboratories, describing and cataloguing their findings. But they discovered that their dredge hauls were disappointingly puny and they soon abandoned it in favor of an ordinary beam trawl. This was a heavy length of wood, like a railway sleeper, to which a lead-weighted net was attached at either end. Being larger and more flexible than the iron dredge, this produced much greater results.

GATEWAY STATION

Gibraltar, Mediterranean Sea, January 18, 1873, 36° 09′ N, 05° 21′ W

In the early hours of January 18, 1873 *Challenger* entered the Mediterranean Sea through the Straits of Gibraltar, slipping quietly past the Pillars of Hercules as a huge gibbous moon climbed silently into the sky from behind the summit of the rock. Herbert Swire was on deck to watch their arrival, woken from an uneasy sleep by the realization that *Challenger* would soon be braving the deep waters of the Atlantic as it finally left the familiar seas of Europe behind. The first leg of their long journey was behind them, the proving leg where they had tested their new technology of dredging and sounding. There was much to be proud of. They had used the "Fox Dipping Circle" to make numerous observations of the strength of the earth's magnetic field. They had used thermally sensitive "galvanic wires" deployed overboard to confirm Wyville Thomson's theory that the deep waters were divided into great discrete masses that moved slowly but with massive momentum about the globe. They had discovered, too, that the muddy bottom of the European continental shelf extended from shore at least 31 miles before dropping steeply into the Atlantic abyss, and they had brought on board a great variety of living organisms dredged from the bottom.

As well as the crinoid that had caused so much excitement among the Scientifics, they had collected a Venus Flower Basket (to which the Scientifics, with their penchant for dog Latin, insisted on referring as *Euplectella*), a conical tube only 2 inches wide at the top, with walls made up of a delicate tracery of tissue that resembled spun glass. Previously, this strange organism had been known only from the deep waters around the Philippines. Together with spectacular finds of strange and grotesque deepwater fish previously unknown to science, their eyes almost blown out of their heads by the thin pressure of the sunlit world, these *Euplectella* confirmed in the minds of *Challenger's* Scientifics that Mr. Darwin had to be correct: the seafloor was a haven for species previously thought to be extinct. The silent landscape was a living portal into the history of life.

By now the portents and ill omens that had dogged the start of their journey were receding quickly into memory. To a man, the crew was pleased to be at Gibraltar. It was, after all, home from home for the citizens of Victoria's empire, the gateway to the east and to the jewel in the crown, as well as the refueling station for voyages to America and the colonies of Africa and the antipodes. The Treaty of Utrecht had ceded Gibraltar to England in 1713 and, as William Spry observed with his customary imperial conceit, it was only since then that Gibraltar's greatest features had been added. These features included the mile-long network of galleries that lined the rock and the honeycomb of tunnels that afforded the garrison protection and access to all points of the compass from which they might be threatened. Joe Matkin noted with amazement that Gibraltar was one of the most heavily armed outposts of the empire. There were no fewer than 1,873 guns on the rock—a surfeit of armament precisely matching the year of their visit. The Channel Fleet, an impressive array of the ironclads *Minotaur*, *Agincourt*, *Sultan*, *Hercules*, *Bellerophon*, and *Lively*, rode at anchor in the Inner Mole, where Swire was keen to catch up with his former shipmates. But the fleet sailed for Madeira later that afternoon, slipping their moorings and steaming Indian-file into the Mediterranean before turning for the straits and the west. Swire consoled himself with the thought that *Challenger* would catch up with them again in Madeira or Tenerife.

And there was much to enjoy. All hands agreed that the view from the signal station at the summit of the rock was magnificent. It was 1,300 feet above sea level and a hard climb along a narrow path, but the effort was amply rewarded by the sight of the glittering waters of the Mediterranean stretching away to the east and the rich ochres and blues of the Spanish hills beyond the cluster of small villages across the bay. To the south, on the other side of the straits, was the imposing dome of Apes Hill and even from the signal station they could see the monkeys clambering among the rocks. Beyond that, fading into the haze was the vast secret bulk of Africa, the source of so much of the empire's wealth. As they stood there, they all felt it, a feeling so intense that it made the hairs on the back of their neck stand up, that particular pride at being an Englishman in the year 1873. It was the height of empire—and it could never fall.

William Spry noted the security arrangements with approval. The daily opening of the gates was carried out with a sense of ceremony appropriate to a lonely British outpost on the periphery of Europe. Immediately after sunrise the Sergeant of the Guard collected the heavy bunch of gate keys from the Governor's house and, accompanied by troops with rifles and fixed bayonets, proceeded to open each gate in turn and lower its drawbridge. Throughout the day the motley collection of nationalities— Englishmen, Spaniards, Portuguese, Turks, Moors, and Jews—that made up the garrison came and went, the Spaniards with compulsory visas, until, at sunset, the ceremony was repeated in reverse, the Sergeant accompanied by his armed guards pulling up the drawbridges and closing and locking the massive gates before returning the keys to the Governor. At night Gibraltar became again a little piece of England, secure from the infidel who waited just on the other side of the gates.

The naval dockyard was small but beautifully equipped, with at least 10,000 tons of coal on hand at any time. An excellent library was immediately placed at the disposal of the crew. It was updated daily with newspapers, periodicals, and telegrams from home. The gardens of Gibraltar afforded many pleasant walks among beautiful surroundings, the colors of fuchsia, oleander, and orange groves mixing with those of Spanish broom, subtropical cacti, and dwarf palm.

But there was work to be done here, too. The Scientifics were excited to discover a large deposit of bones, shells, and teeth in the cave of St. Martin halfway up the southeastern face of the rock and speculated that they might be the remnants of animals that had lived there during the last great glacial age. Samples were taken for further work on board the ship. While they were at Gibraltar, the naval men made a new survey of the Mole, rated the chronometers, and calibrated their compasses. The distance to Malta was calculated too, by measuring the time taken for an electrical impulse to transit the newly laid telegraph cable between the two colonies, and found to be exactly 1,000 miles. Coincidentally, the ship in Malta that assisted in this measurement was the *Shearwater*, the same sloop that Wyville Thomson and William Carpenter had used for their survey of the Mediterranean in 1870. The disappointment of that mission was at least partly responsible for the decision that *Challenger* would not enter any further into the Mediterranean, thereby missing one of the most extraordinary stories that the silent landscape has to tell.

THE DESERT UNDER THE SEA

That story had to wait another hundred years for new technology as well as a new ship. The new technology was seismic shooting and the new ship was the spiritual successor to, as well as the namesake of the original *Challenger*, the drilling vessel *GLOMAR* (*GLObal MARine*) *Challenger*.

At about the same time that multibeam sonar was invented, scientists were developing another method of seafloor imaging using sound—seismic shooting. Seismic shooting was pioneered by the American geophysicist Maurice "Doc" Ewing, the founder of the Lamont-Doherty Geological Observatory in Palisades, New York, an institution that became the home of marine geology in the United States. Ewing's idea was to use dynamite to create an explosion so powerful that the sound waves were not just reflected back from the ocean floor but penetrated it and were then reflected back by subsurface features. With suitably positioned detectors, the time taken for the echo of the explosion to return to the surface indicated the hardness of the sedimentary layers beneath. The American research vessel *Chain*, operating out of the Woods Hole Oceano-graphic Institution in the late 1950s, used this technique while sur-veying in the Mediterranean. In the course of that cruise, seismic shooting revealed the presence of a hard layer—an acoustic reflec-tor or sound-reflecting layer—deep within the sediments of the Mediterranean seafloor. This M-reflector, as *Chain's* senior scientist Brackett Hersey named it (M for Mediterranean), was quite clearly something that had been laid down with the sediments of the seaf-loor because it followed perfectly the contours of the hard rock basement of the Mediterranean.

Although sonar gave a clue that something strange had hap-pened in the Mediterranean, a physical investigation of the nature of the hard layer was still very far off. Because the M-reflector was buried under 2 kilometers of sediment, the only way to reach it was by drilling and that meant using a drilling ship. Only one vessel was suited for the job, the research vessel of the Deep Sea Drilling Project, the *GLOMAR Challenger*. The origins of the Deep Sea Drilling Project are complex and will be dealt with in the final chapter of this volume, when we consider the implications and heritage of the original *Challenger* expedition. However, since we cannot leave the Mediterranean region without discussing its secret history we must anticipate ourselves slightly and tell the story of the day the Mediterranean became a desert.

In late August 1970, the *GLOMAR Challenger*, after 12 success-ful cruises around the Atlantic, entered the Mediterranean Sea for the first time. Interest in the Mediterranean was intense for two reasons: first, because the new theory of plate tectonics predicted that the proto-Mediterranean must once have been a wider ocean—named Tethys after the world-girdling ocean of Greek mythology—which had gradually narrowed as the African plate moved northward into Europe; and second, because scientists still wanted to know what the enigmatic M-layer was. Indeed Bill Ryan, one of the *GLOMAR Challenger's* two chief scientists on its Mediterranean cruise, had been with Brackett Hersey aboard the *Chain* when the M-layer was discovered. But by the evening of August 24, 1970, Ryan and the other chief scientist aboard the *GLOMAR Challenger*, Ken Hsu, were becoming discouraged. They had drilled down to the M-layer and all that they had managed to retrieve was a handful of gravel. Late that night, as they sat mournfully in the deserted paleontology lab, Hsu watched in growing amazement as Ryan washed and rinsed the gravels that they had retrieved, dried them on a hotplate, and started gluing them to the cover of a brown manila folder.

But as he watched, Hsu gradually became aware of the reason for his colleague's apparently compulsive preoccupation. Ryan was performing a form of low-tech grain-size analysis; he was ordering the sediment grains according to their size in order to better understand the forces that had laid them down. When he was finished, both scientists could see that the larger grains were fully 7 millimeters across; genuinely gravel sized and not at all something that either had expected to find in the deep ocean. Gravels are normally deposited by fast-flowing rivers and as they are too large to be moved by the slow-flowing currents of the deep ocean, they tend to settle out close to the mouths of estuaries. Among the gravels there were shiny crystals of the mineral gypsum, the residue left behind when seawater evaporates. Today gypsum forms on the shores of arid coasts and is also found in rock outcrops on land that were formed under conditions of intense evaporation; it belongs to a class of minerals known as evaporites. No one expected to find gypsum in a deep-sea deposit, much less in association with gravel-sized grains. At its current position the *GLOMAR Challenger* was drilling on the flank of a submerged volcano, but it was clear that the volcano could not have provided these pea-sized gravels unless it had once been exposed to weather and subjected to erosion by streams running down its sides.

Then the paleontologists on board began telling Hsu and Ryan that the species of foraminifera—tiny single-celled creatures that leave a hard shell of chalk—that they were retrieving belonged to contemporary species that are usually found only in shallow coastal lagoons. In the deep waters where they were now drilling this was a very strange finding. But these "forams" had another unusual feature: They were dwarfs. When dwarf faunas are found in the fossil record, they are invariably associated with times of environmental stress. To Hsu there could be only one explanation for the gypsum, gravel, and stressed foram fauna at the bottom of a hole drilled in 3 kilometers of water: The Mediterranean basin had once been a desert.

Hsu was able to piece together the story. At some point in the past, the supply of seawater from the Atlantic Ocean through the Straits of Gibraltar had stopped. As it dried out, the Mediterranean basin gradually changed into a giant salt lake, like the Dead Sea but a hundred times larger. As the desiccation continued the remaining brine became progressively saltier in the harsh Mediterranean sun, stressing the foraminifera and other deep-dwelling creatures that lived there and turning each new generation into a progressively more stunted shadow of its ancestors. Finally, the Mediterranean bottom had been laid bare and the submarine volcano near which the ship currently floated became a mountain, rising high above a barren treeless landscape. Its oozes and sediments were baked hard by the sun and eventually eroded into gravel by the streams running down its flanks. Finally the Straits of Gibraltar had opened again and the Atlantic had come cascading into the 3-kilometer-deep ocean basin in a torrent that dwarfed even the Victoria Falls on the Nile.

Ryan was skeptical, and who could blame him? Hsu had fabricated this extraordinary story in the course of a few days on the basis of three lines of evidence that, on their own, were extremely flimsy. But Hsu was not discouraged. He pointed out to Ryan that the M-layer itself could well be additional supporting evidence for the desiccation of the Mediterranean. Did not the M-layer follow the Med's basement topography perfectly? Did not the seismic data show unequivocally that sediments must underlie, as well as overlie, the layer? The M-layer must, therefore have been deposited when the basin topography of the Med had already been formed. It was the hard relic of the ancient saltpan.

A couple of holes later they found more supporting evidence for Hsu's theory. At Hole 124 they discovered rocks that Hsu, with characteristic bravado, named the "Pillars of Atlantis." These cores were composed of the mineral anhydrite and contained a type of fossil known as stromatolite. Anhydrite is found only on arid coastal flats in a dry environment known as sabhka. Stromatolites are hard, limey, fist-sized fossils secreted by billions of tiny, filamentous, blue-green algae. These tiny bacteria-shaped objects, which have no true nucleus, live today in shallow salty pools at the edges of oceans, for example on the margins of Shark Bay in Western Australia. As the tide comes and goes, they trap limey muds between the layers of cells that eventually become lithified—turn into rock—and what was once a gelatinous mass of primitive cells becomes a large hard mound. Stromatolites are one of the oldest fossils in the geological record. They have been found in rocks that are 3.5 *billion* years old (about 80 percent as old as Earth itself). But the point is that the stromatolites are *always* found in shallow waters. They need sunlight to power the process by which they derive their nutrition: photosynthesis.

Ryan began to waver. The presence of anhydrite and stromatolite in this core was indeed strong confirming evidence. This left the remaining crucial question—when? Not all members of the crew became converts to Hsu's apocalyptic vision and indeed, when the ship eventually returned to Lisbon after 60 days at sea, only three people were willing to put their names to the paper that proposed the desiccation idea: Hsu, Ryan (who had finally been persuaded by the weight of evidence), and the foram paleontologist on board, Maria-Bianca Cita. Cita's contribution to the desiccation idea was most notably the dating of the event. She pointed out that many of the lands rimming the Mediterranean—Italy, Spain, Portugal, and several of the North African states—have extensive deposits of salt. When dated using land fossils, all of these deposits were found to be of the same age, defining a short-lived but discrete period at the top of the Miocene epoch named (after the location of the biggest salt deposit at Messina in Sicily) the Messinian. The forams that Cita analyzed on the cruise had all been of Messinian age and thus pinpointed the age of the M-layer very precisely between 5 and 6 million years old. More than a century before, English geologist Sir Charles Lyell had differentiated between the Miocene and the Pliocene epochs of Earth history on the basis of two radically different fossil assemblages, the changeover occurring within this narrow Messinian stage.

In the late 1960s and early 1970s deep-sea drilling technology was still very new, and strangely, one of the most difficult rock types to retrieve was soft salt. Its solubility meant that it was almost invariably dissolved away by the circulating sea water used to keep the drill bit cool and lubricated. Amazingly, at the very last hole they drilled before their return to Lisbon and against all the odds, the scientific party retrieved a cylinder of salt. It was from the Messinian stage. The enigmatic M-layer could now be finally understood: it was indeed the layer of evaporite that defined the end of the Messinian salinity crisis.

When the ship docked in Lisbon, the crew was unprepared for the publicity that was beginning to surround Hsu's big idea. It had already made the front pages of dailies around the world and the co-chiefs of the expedition were bombarded with requests for interviews. The idea that the huge Mediterranean Sea had once been as dry as a garden pond after a long hot summer had caught the public imagination. It was good publicity too for the Deep Sea Drilling Project, whose success in proving the plate tectonic theory had made the headlines only a year before.

Hsu soon found that the scientific literature was dotted with many other clues to the desiccation idea. For example, previous seismic surveys revealed strange elongated trenches under the Mediterranean seafloor. Some geologists speculated that these trenches were river channels that had submerged when a chunk of the earth's crust had sunk, but only one person, a young French scientist, thought that they might be the remains of river beds incised into the edges of the Mediterranean when that sea had dried out millennia ago and the rivers draining Europe and North Africa had cut down to its bottom.

Still more impressive was evidence from the mouths of the Rhone and the Nile rivers. At the end of the nineteenth century the remnants of a deep gorge had been found beneath the Rhone. This gorge had been filled in with sediments during the Pliocene Epoch and then overlain by contemporary sands and gravels. Further exploration showed that this fossilized gorge extended 200 kilometers from Lyon to La Camargue, practically to the shores of the Mediterranean, where its bedrock base was now a kilometer below present-day sea level. The conclusion was inescapable: The mighty Rhone had once been even mightier, cutting down through a thousand meters of granite to deliver its cargo of water to the desiccated floor of the Miocene Mediterranean.

Similarly, Soviet geologists, surveying prior to the construction of the Aswan Dam, had drilled several boreholes into the floor of the Nile Valley and to their amazement had discovered another deep narrow gorge similarly filled with Pliocene sediments. It was the mirror image—on the southern margin of the Mediterranean—of what had been found beneath the Rhone: a deep channel where the waters of the Nile had once flowed into the Mediterranean with much greater ferocity than today. Other buried gorges have since been discovered in Algeria, Israel, Syria, and other countries, all adding silent support to this most amazing of tall sea tales.

But the biggest irony of all came when Hsu discovered that one Herbert George Wells, too, had believed that the Mediterranean was once a desert. H. G. Wells studied geology under Vincent Illing at Imperial College in London before the First World War. Finding Lyell's account of the faunal crisis separating the Miocene and Pliocene epochs in the library one day, Wells later used it as the basis for one of his science fiction novellas (*The Grisly Folk*, 1921) in which he speculated that the Mediterranean had once been a massive, dried-out hole separating Europe and Africa.

Whoever said that truth is stranger than fiction?

The Restless Earth

Gibraltar, Mediterranean Sea, January 23, 1873, 36° 09' N, 05° 21' W
to Station 19, Western Atlantic, March 3, 1873, 19° 30' N, 57° 35' W

ATLANTIC TRANSECT

HMS *Challenger* left Gibraltar on January 23, 1873, and finally
headed out into the true unknown, the deep Atlantic Ocean. As the
continental shelf dropped away beneath her, and the fans of mud
from the rivers of Northern Europe gave way to the unchanging
vastness of the Atlantic abyssal plain, the crew settled into a routine
of sounding and dredging operations that would govern all their
days for the rest of the voyage. The ship's company rose at 4 A.M.
and, on standing orders from *Challenger's* two staff surgeons, swabbed
the decks with copious quantities of seawater. The emphasis on
hygiene, like the daily ration of lime juice and the improvements in
discipline were all part of the navy's new recognition of the
importance of morale. And although the bluejackets might com-
plain at the extra work involved in the daily swabbing of the decks,
they, too, recognized its importance and so they did it well. Legends
of the navy only 60 years before, when impressment was still the
norm and men did not see their families for years on end, as well as
the terrible privations suffered by sailors during the Crimean War,
were still too fresh for the new sail-and-steam navy not to appreciate the reforms.

The washing down of *Challenger* was followed by the tidying of the sleeping quarters; hammocks, sheets, and blankets were all stowed neatly. At 6 A.M. there was a simple breakfast of biscuit and cocoa and then all hands started preparations for the day's dredging and sounding operations, furling the sails while William Spry and his mates made steam down below in the engine room. It was essential to use steam power while dredging and sounding, because only with a consistent push from the propeller could the attitude and position of the ship be kept constant. Under sail she would drift miles off-station, but with the steam engine churning and her bow head-on into the sea, enough way could be put on to achieve a stable position from which the sounding line—a rope of woven piano wire—could be run out. But sounding by *Challenger*, unlike most conventional sounding aboard Victorian men o'war, was more than just an exercise for determining the depth of the ocean. Attached to the sounding wire were various instruments by which the silent landscape—and its watery atmosphere—could be probed. There were three main types of instrument: thermometers for measuring the water temperature at the seafloor (as well as at various depths in the ocean), specially constructed flasks for taking samples of deep and bottom waters, and a device for retrieving sediment from the seafloor itself.

All of these instruments were marvels of Victorian ingenuity and the thermometers and water samplers were just as extraordinary as the sampling devices described in the first chapter of this book. The thermometers were of the maximum and minimum kind originally invented by James Six in the eighteenth century and are still known by that name today. They were used to measure the highest and lowest temperatures encountered in the long drop to the sea bottom and were in all respects identical to those used today in garden greenhouses, except that they were heavily armored against the crushing pressure of the deep ocean. They consisted of a curved U-tube filled with mercury attached to a bulb containing creosote.

The expansion or contraction of the creosote moved the curved U of mercury around in the two arms, and as each limb of the mercury column moved, it pushed a small spring-loaded metal index in front of it. The base of each index—that is, the end nearest the mercury—then stayed in place as the mercury moved away again. Thus, the highest and lowest temperatures were preserved for *Challenger's* scientists to log when the thermometer was hauled up again.

The water-sampling flasks were of two varieties: one constructed to sample bottom waters and the other to take samples at different depths. The bottom-water sampler was a slender, reinforced rod with finely machined discs of brass at either end. A sleeve above the upper disc was attached to a lanyard so that when the bottom was finally reached, the lanyard disengaged a clutch that allowed the sleeve to slide down the rod, trapping a sample of bottom water between the two discs. The intermediate water sampler was similarly constructed but had two valved stopcocks at top and bottom through which water flowed freely as the device descended, but which snapped closed as soon as the line was hauled upward. In this way these devices could be lowered to a predetermined depth and then used to trap water from that depth.

The thermometers had their limitations. Although the bulbs supposedly compensated for variations in pressure by enclosure in a second bulb full of alcohol, the accuracy of the pressure compensation was something of an unknown. Also, the long journey through the water column and the emerging realization that temperature did not decrease in a straightforward manner with depth meant that the maximum and minimum temperatures registered were not necessarily those of surface waters and sea bottom, respectively. To minimize this problem it became routine aboard *Challenger* to deploy the line again after the initial sounding, this time with thermometers arrayed every hundred fathoms to try to chart more accurately the change of temperature with depth.

The whole complex apparatus was rounded off by the sediment sampling apparatus that dangled at the bottom of the rope and was supposed to retrieve a sample of the seafloor for later study aboard the ship. Even this technique was not straightforward, because the crew had to know when the sounding line had reached bottom and once the line was in the water there was no way of telling whether it was vertical beneath the ship's keel or drifting off at an unknown angle under the influence of some deepwater current. The answer to this problem was simplicity itself. The sounding wire was calibrated in increments of 25 fathoms: 100-fathom marks being blue, 50-fathom marks red, and the 25- and 75-fathom marks white. As the line was fed out, the time taken for successive markers to pass over the side increased proportional to the drag of the line. When the time taken for the fathom markers to pay out departed from this precise relationship—the winchman had a book of tables that informed him—the bottom of the ocean had been reached. The crew then pulled in the line, carefully read and noted the thermometer readings, sent the water flask down to Buchanan's physical laboratory where its specific gravity and other physical properties were measured, and sent the contents of the sounding rod to the natural history laboratory where the composition of the ocean floor was noted and the sediment or rock sample dried and bottled. Figure 4 shows the *Challenger* sounding and dredging apparatus. Then, the second deployment—to measure the temperature of the ocean at different depths—was tackled. And all this before the day's quota of dredging action had even started!

At last the Scientifics were in their element! During the first days out of Portsmouth they had found, as William Spry reported, "the etiquette and usages of naval everyday life . . . particularly vexatious and annoying" but with the subtropical weather now set fair and some real science to do, they were at last engaged in what they had come for. But for the crew and the ratings aboard the ship

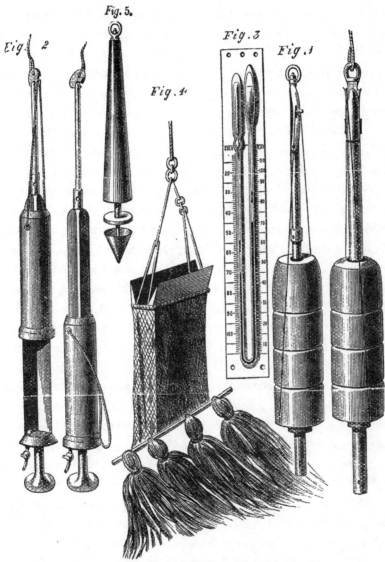

Fig. 5.

Fig. 2

Fig. 4.

Fig. 3

Fig. 1

SOUNDING AND DREDGING APPARATUS.

Fig. 1. Sounding-machines. Fig. 2. Slip water-bottle. Fig. 3. Deep-sea
thermometer. Fig. 4. The dredge. Fig. 5. Cup sounding-lead.

FIGURE 4 Sounding and dredging apparatus

life was not quite as sunny. The constant stop-and-measure of the *Challenger* expedition was quite at variance with what they were used to as military men when the emphasis was on arriving at one's destination with dispatch. Even Captain Nares who, as an Arctic explorer, was accustomed to the methods of science, must have found the drudgery wearisome.

Joe Matkin was getting used to his quarters below decks in an issuing room that was not nearly as comfortable as the one that he had enjoyed aboard *Audacious*, because on *Challenger* it was below the water line and suffered unbearably from the poor ventilation. Also it was so small that there was no room for a seat, so he had to write his letters standing up. These hardships were an additional burden to a young man who was already worried about his father's failing health—and it didn't help either that in Madeira he had witnessed no less a personage than the Mayor of Edinburgh comfortably overwintering in the subtropical balm for the sake of his own sickly constitution.

The expedition plan called for a transect of the deep Atlantic, with attendant temperature readings and seawater sampling, from Santa Cruz in Tenerife to Sombrero in the Virgin Islands. They planned to take soundings every 120 miles and dredge every 300 miles, so a total of 22 soundings and 13 dredgings were forecast for the entire 2,700-mile trip across the Atlantic. In preparation, *Challenger* took on 25 tons of coal at Santa Cruz. This port was where Nelson lost his eye but, as Joe Matkin wryly reflected, a modern ironclad could knock the place to ashes in about two minutes. And it was at Santa Cruz that death very nearly visited *Challenger* again: One of the boys fell from the rigging, a height of almost 40 feet into the sea, and but for the quick action of an able seaman he would surely have drowned.

On the evening of Friday, February 14, *Challenger* left Santa Cruz. The weather, notes the expedition report, was "bright and pleasant, with a light breeze blowing from the northeast." For the first 250 miles west of Tenerife the ocean bottom was found to be nearly level at a depth of about 2,000 fathoms, consisting for the most part only of Globigerina ooze (of which more later) but only 50 miles further to the west the depth of the ocean shoaled suddenly to only 1,500 fathoms and the nature of the ocean bottom changed dramatically.

METAL PEARLS OF THE DEEP

On Tuesday, February 18, 1873 *Challenger* hove to some 160 miles southwest of the tiny island of Ferro (part of the Canaries) at a site designated by Wyville Thomson as Station 3 of the Atlantic transect. The soundings indicated a modest depth of only 525 fathoms and there was no expectation of anything out of the ordinary when the dredge was sent down at 10 o'clock that morning. As always, the dredge took an age to reach the seafloor and, after the usual several hours of towing under steam, yet another age before it reached the surface again, finally arriving back at 5 o'clock in the afternoon. It was just another routine chore in what, it was becoming clear, was going to be a long series of such chores and more than one bluejacket must have already been wondering whether signing up for a three-and-a-half-year voyage of scientific discovery had been such a good idea after all. So none of them could have guessed that they were about to make one of the most momentous discoveries of the long voyage, one that would have important implications for an industrial society 140 years in the future.

When, with the evening sun already beginning to dip toward the western horizon, the dredge finally surfaced, all on board could see that within the mesh were fragments of corals that Wyville Thomson identified as members of the *Corallium* genus. Despite the fact that the corals were quite dead, amazingly they all gleamed with a peculiar blackish metallic luster, and at the base of each piece of coral were attached fractured lumps of brown-black rock showing clear signs of concentric laminations. Buchanan's chemical tests showed that the peculiar luster of the dead corals was caused by a coating of peroxide of manganese, while the blackish lumps to which they attached were apparently some form of phosphate secretion. As the ship moved on across the vast expanse of the Atlantic, Buchanan labored in his small chemical laboratory, racking his brain as he tried experiment after experiment to try to refine his assessment of these fragments.

Even Lord George Campbell momentarily forgot his lack of interest in science as he, too, was caught up in the excitement. His journal records that ". . . 160 miles to the southwest of the Canaries we found a shallow bank of 1,525 fathoms; having the day before sounded in nearly 2,000 fathoms and dredged there getting half a bag full of globigerina mud and that was all. The dredge was sent down on this bank and came up with a quantity of large dead coral, coated with black stuff, while on the branches beautiful large sponges of a new species were growing. A number of small white starfish also came up, but no mud, so the dredge must have scraped over a forest of dead coral."

Some 1,900 miles to the west they came across another field of the strange metallic fragments. But here was a difference, because outnumbering the fragments by thousands to one were the objects from which they obviously must have originated, potato-shaped nodules ranging from walnut-size to grapefruit-size. They were strewn across the seafloor of the western Atlantic in their billions. It was incredible, it was beyond belief. What were these strange metal potatoes of the deep?

And it was at this point—some 600 miles to the east of their destination in the Danish West Indies at Station 16 of the first Atlantic transect—that the Scientifics made the connection between the black patina on the Station 3 corals and the fragments attached to their bases. Buchanan discovered that the phosphate secretions from there were as rich in manganese as the patina on the Station 3 coral fragments. The two were related—there was something in the deep that was concentrating this rarest of metals. The Scientifics also noted that regardless of size, these nodules all exhibited a characteristic concentric zoning when cut in half that reminded them of how trees lay down their annual growth layers. From there it was but a small mental step to wondering whether this zoning might reflect some similar process with metals precipitating out of seawater over the course of millennia to form the concretions, perhaps akin to the way a pearl gradually builds up in layers around a tiny speck of grit within an oyster shell. "The whole question," wrote Wyville Thomson in his account for *Nature*, "is a very singular one."

Challenger's Scientifics had discovered the first of what was destined to become a common sight for them as they made their laborious way around the world, balls of metal-rich rock that today go by the common, if inaccurate, name of "manganese nodules." The expedition was to encounter them not only on the east and west sides of the mid-Atlantic ridge but also in the North Pacific between Japan and the Sandwich Islands (Hawaii), in the vicinity of the Sandwich Islands themselves, in their biggest find of all, a 3,000-mile arc of them across the South Pacific from Tahiti to Valparaiso.

But at the time of *Challenger's* first encounter with them in the Atlantic during that February and March of 1873, the concretions were just another scientific curiosity. There was no inkling on board that within a hundred years these strange discoveries would excite tremendous curiosity from global mining conglomerates. There was no inkling either that *Challenger's* final find in the South Pacific would stimulate the financial juices of one of the biggest industrial predators of the twentieth century, Howard Hughes.

Today we recognize that the phrase "manganese nodule" is something of a misnomer because there is more than just manganese in the nodules. A typical nodule contains about one-third manganese with a second significant component (about one-twentieth of the total mass) being iron. The rest of the nodule is composed of variable amounts of copper and nickel and there are trace elements of cobalt, titanium, and aluminum. The concretions tend to come in one of three distinct shapes: the single cored nodule, known as the mononodule; nodules with several cores, called polynodules; and the composite nodule where several nodules are joined together.

Without exception, and as Wyville Thomson and Buchanan originally discovered that long ago day on *Challenger*, the heart of each nodular core is a tiny fragment of sea-bed detritus, be it a shark's tooth, a pebble, or a tiny fragment of volcanic pumice. Their analogy with an oyster's pearl was close to the truth. But while it is well known that the genesis of the pearl is the gradual accretion of sticky mucus around an irritating piece of grit in an oyster's shell, even today there is still no consensus as to how the metallic nodules of the deep sea accrete around their own nucleus. Wyville Thomson was convinced that the buildup of the metallic layers was related to some form of biological activity. "The manganese," said he, "is doubtless, like the iron, set free by the decomposition of . . . organic bodies and tests" and even today there are those who think that the process by which the nodules form is totally organic. A more popular class of theories however, revolves around the inorganic formation of polymetallic nodules. These theories include the slow precipitation of dissolved metals from seawater catalyzed by some as yet unknown process; the rapid deposition of the nodule's layers in ocean regions where high temperatures and concentrations of dissolved metals abound (for example, near hydrothermal vent systems; of which more later), or by some form of remobilization process by which metals buried in sediments are re-precipitated as nodules.

Interestingly, even within this "inorganic" class of theories the idea that some form of biological mediation is involved in nodule formation continues. As long ago as 1928, the Soviet scientist V. S. Butkevich reported the discovery of bacteria within polymetallic nodules from the Arctic basin. It now seems likely that there is more than one way to precipitate metallic nodules in the deep sea but, as Wyville Thomson recognized, there is one thing that all concentrations of nodules have in common: they are usually found in massive fields on the sea bed. It was *Challenger's* Scientifics who discovered that these fields of metallic nodules are always associated with a type of sea bottom that they called "red-clay," and, as we shall see in the next chapter, it was the search for an explanation for this type of sea bottom that turned up yet another extraordinary facet of the lore of the sea—the fact that deep waters are acid.

But as *Challenger* approached the West Indies the crew encountered a phenomenon that dwarfed in importance even the discovery of manganese nodules. At Station 19, just four days later, they plumbed extraordinary new depths and, as they drew up the depth transects, saw that they had confirmed the existence of a mid-Atlantic plateau. What they did not know was that in so doing they had unwittingly set the stage for the advent of a completely new science.

THE WOUND THAT NEVER HEALS

The Scientifics had confirmed what the American oceanographer Maury had already suspected from his primitive, low-resolution soundings some years before, that there was a topographic high, which Maury had named the "Dolphin Rise," in the middle of the Atlantic. Future transects would show, before *Challenger* returned to England, that this topographic high is part of a chain of submerged mountains running down the middle of the Atlantic from the Arctic to beyond the tip of the Cape of Good Hope. Today you can see this very clearly on topographic maps of the ocean floor, which can be found in every university geology department in the world.

I first came across such maps as a graduate student at Cambridge, England in the mid-1980s and I can remember being amazed at their clarity—amazed and also perplexed, because I had no idea that the ocean floor was sufficiently well surveyed to show this level of detail.

Challenger's confirmation of the existence of a mid-Atlantic ridge would, within a hundred years, have revolutionary implications for the science of geology by substantiating claims of the theory of plate tectonics. As long ago as the seventeenth century, the philosopher Francis Bacon noted the apparent fit of the east coast of South America into the west coast of West Africa, and speculated that the two had once been joined together. But it was the German meteorologist Alfred Wegener who, while lying in a hospital bed during the First World War, formalized this observation into a proper theory. Wegener noticed not only the fit of South America and Africa, but also the apparent continuation of rock strata in Brazil's Amazon basin with strata in West Africa despite their separation by the wide Atlantic Ocean. Furthermore, the geology of the Appalachian Mountain chain of North America seemed curiously similar to that of the highlands of Scotland; also the Paleozoic fossils of North America and Scotland—not to mention many other places around the globe—seemed curiously similar, too. How could this similarity be explained if populations of these fossils had been separated by oceans for all eternity? It was these clues that led Wegener to propose that the present-day continents must have—some 200 million year ago—been fused into a single supercontinent.

He put the idea forward in a book, *The Origin of Continents and Oceans*, which was published in 1915. Wegener argued that as this single supercontinent (that he named Pangea) broke apart, the resulting fragments slowly traveled to their present positions, creating the ocean basins in the process. Wegener's theory, which came to be known as "continental drift," was immediately and loudly derided by the geological community, which was perplexing, because at least his theory offered a neat mechanism to explain the fit of the continents and the strange spread of fossil communities on either side of supposedly immutable ocean basins. Any other explanation would have to invoke some very convoluted logic, including the formation of temporary land bridges, which then vanished without trace from the geological record, as well as complicated, sudden, and massive vertical motions of entire continents that would have had them going up and down like the pistons of *Challenger's* steam engine. These explanations became progressively more elaborate and unlikely with every strange new fossil discovered.

Wegener's theory of continental drift explained not only why the continents seemed to fit together so well and why similar fossil communities could be found so far apart, but also the evidence of glaciation (such as glacial sediments and hanging valleys) in regions of the Earth (such as the equator) that quite patently could never have been covered with ice if the continents had been static. Yet Wegener was reviled for his theory, often in quite personal terms, because he simply could not provide a *mechanism* by which the continents could move. Critics argued that the passage of these massive blocks of rock through a supposedly solid seafloor would have ground them to fragments before they traveled any significant distance.

And there the matter rested. Continental drift was a one-hit wonder, a theory that looked superficially plausible but that, without a mechanism, was little better than idle speculation. In America particularly, the musings of Wegener were widely scorned. Indeed, to dally with the theory publicly could do serious damage to a geologist's career. And yet echo-sounding maps of submarine topography continued to accumulate.

Bruce Heezen of the Lamont-Doherty Geological Observatory at Palisades, New York, was deeply involved in the generation of those echo-sounding maps. He had been recruited from the University of Iowa by Maurice Ewing. Heezen was starting to combine his own hard-won seismic data with pre-existing data from the German *Meteor* cruise and even the voyage of USS *Stewart*, where Hayes had first field-tested the prototype of his sonic depth finder. By this time Heezen had hired himself a technician named Marie Tharp who specialized in drawing maps. It was the combination of her peerless drafting skills and ability to visualize complex shapes in three dimensions with Heezen's talent for synthesizing geological data that culminated in the production of some of the most important pieces of scientific art of the twentieth century: the supremely detailed maps of the ocean floor mentioned earlier.

Heezen laboriously constructed six seismic profiles across the North Atlantic Ocean by combining records from the cruises of the three vessels. At that time, the Cold War was at its height and the

U.S. Navy generally took a dim view of any civilians—even scientists—who took it upon themselves to construct bathymetric contour maps. These maps were the basis by which their own—and Soviet—nuclear submarines navigated and, as such, this information was considered highly confidential. Despite naval funding, therefore, Heezen and Tharp hit upon the idea of drawing perspective maps: three-dimensional dioramas of the seafloor with enough detail to show the mountains of the mid-Atlantic ridge but not enough to allow a Soviet sub to sneak up on the eastern seaboard of the United States and vaporize New York. These maps eventually became so beautifully colored and intricate that *National Geographic* magazine published them as soon as they left Tharp's drawing board. The first of Tharp's maps—of the North Atlantic—showed a scene of startling symmetry that the *Challenger's* Scientifics would have given much to see. This showed the steep slope of the American continental shelf, the western Atlantic abyssal plain rising, slowly at first then more and more sharply, up to the summit of the mid-Atlantic ridge, the descent again into the eastern Atlantic abyssal plain, and then the final shoaling up onto the Afro-European continental shelf. But it also showed something else. At the summit of the mid-Atlantic ridge it showed a groove or notch where the summits of the peaks should have been. Heezen took one look before dismissing it as an artifact of the drafting technique. He knew immediately what the implication of the notch was and didn't feel like taking risks with his glittering career. The groove resembled the rifts between continental blocks that Wegener's theory of continental drift predicted far too closely for comfort.

Also at that time, the geology of a vast valley running between the Red Sea and Lake Turkana in East Africa was just beginning to be understood. Most geologists agreed that this was an area where the Earth's crust was splitting. In fact, the block of continental granite that made up the floor of the valley was subsiding into the Earth's crust as the sides of the valley pulled apart. For this reason the valley was known as the East African Rift Valley. When Tharp took the height transects across the East African Rift Valley and transformed them into one of her topographic maps, they looked very similar to the notched profile across the mid-Atlantic ridge. It was enough to give Heezen pause.

The clincher came through the mapping of the seismic waves made by earthquakes. Seismologists Charles Richter and Beno Gutenberg had, in the 1940s and 1950s, determined, by using seismic triangulation techniques, that the epicenters of Atlantic submarine earthquakes lay in a line seemingly close to that of the mid-Atlantic ridge. When Heezen had Tharp plot them on the map she was drafting of the Atlantic, he found that they lay in a line *precisely* down the groove that he and Tharp had discovered at the center of the ridge. Now the conclusion was obvious and inescapable: The notch was exactly like the Rift Valley of East Africa, a place where two portions of the Earth's crust were pulling apart and generating Richter and Gutenberg's earthquakes in the process.

But if one accepted this correlation, then one was obliged to take it further, for Richter and Gutenberg's data were not confined to the Atlantic. They had good data for the Indian Ocean and even some for the Pacific that showed that earthquake epicenters were clustered into well-defined lines there, too. And like the Atlantic, in the Indian Ocean the line seemed to be more or less precisely positioned in the middle of the sea (in the Pacific the line of earthquakes is offset for reasons that we shall come to later).

Heezen made a bold intuitive leap. Maybe there were rift valleys in these oceans, too, and, because the line of earthquakes was apparently continuous from the Atlantic into the Indian and then into the Pacific Ocean, maybe the rift valley was continuous, too. In fact, maybe this was a single rift valley that extended all the way around the world under the ocean? It was an idea that immediately generated an unequivocal and testable hypothesis. This "wound-that-never-heals" as Heezen had started thinking of it, if it did indeed extend around the world, should also exist in parts of the ocean that had never been sounded by physical or seismic means. All that was necessary to test that hypothesis was to get on a ship and go to one of these uncharted areas and see. And that is what he did.

Eventually Heezen and Ewing traced the "wound-that-never-heals" from its outcrop on land in Iceland down through the North and South Atlantic, round into the Indian Ocean where it forms the Carlsberg Ridge (which HMS *Challenger* also discovered) and then onto the place where it eventually splits in two, forming the Red Sea on one hand and the East African Rift Valley itself on the other. Another spur passes into the Pacific. The Pacific is older than the Atlantic and shrinking as the Atlantic widens. This is why its mid-ocean ridge is offset toward the west coasts of South and North America. Eventually the "wound-that-never-heals" disappears somewhere in eastern Siberia.

Of course the success of this reconnaissance immediately generated a new set of questions, the most important of which was: If new crust is continually being manufactured at the center of the oceans, what's happening to the old crust? To Heezen this was not a problem; it didn't go anywhere, and as a result the Earth was continually expanding in size. At some point in the deep past the Earth had been completely girdled by continental crust and as the plate tectonic fires raged across millennia and the ocean basins formed, this granite corset had split asunder with the various blocks moving away from each other. The Earth was simply getting larger like an inflating balloon. But Heezen was wrong, and the evidence of that would come from the Pacific.

And yet back in the late nineteenth century, the Pacific was an ocean that HMS *Challenger* would not reach for many months yet. But still there was exciting science to be done in the Atlantic and, as the little corvette slipped quietly to her mooring off Charlotte Amalia, capital of the tiny Caribbean island of St. Thomas, the biggest question was why they were sailing on acid.

Kingdoms of Mud and Lime

Station 19, Western Atlantic, March 11, 1873, 19° 30′ N, 57° 35′ W
to Charlotte Amalie, St Thomas, Danish West Indies, March 16, 1873,
18° 22′ N, 64° 56′ W

THIS CORROSION

Three weeks before, on February 26, 1873, *Challenger* had encountered the greatest depth of their journey so far. At Station 9, latitude
35° 10′ W, longitude 23° 23′ N, bottom was touched at a staggering 3,150 fathoms (almost 19,000 feet). And when the dredge was
hauled in the crew had their first glimpse of a completely new kind
of seafloor sediment, one that was not ooze but rather a monotonous fine-grained silt with a peculiar reddish-brown color. The
Scientifics named this silt "red clay." Lord George Campbell found
himself strangely fascinated by this new secret of the silent landscape, because he had already had enough of ooze. The mess it
made offended his military mind. "The mud!" he wrote "Ye gods.
Imagine a cart full of whitish mud, filled with the minutest shells,
poured all wet and sticky and slimy onto some clean planks and you
may have some faint idea of what globigerina mud is like."

Wyville Thomson was fascinated, too, one of his special interests
being the geographic distribution of different seafloor sediments.
Just after they left the Canaries he noticed that the sediments they
were bringing up were the fine white muds that they had named
"Pteropod" and "Globigerina" ooze. Pteropod ooze was composed

largely of the small white shells of mollusks, Globigerina ooze of the even smaller shells of foraminifera. All on board were amazed that the sea bottom should be such a fecund haven for life—it seemed the final annihilation of Forbes's azoic theory—although there was much debate as to how all these oozes could grow in the lightless depths of the abyss. There was only one man on board who harbored private doubts that the muds were native to the seafloor, and it would be many months before his theories came to fruition in the icy wastes of the Southern Ocean.

As *Challenger* traveled westward across the Atlantic and the soundings became deeper and deeper, the Scientifics found that the white ooze they had become accustomed to finding near the edges of the continental slopes was giving way gradually to gray ooze. Deeper still, and red clay replaced the oozes. Eventually the Scientifics were able to predict, with a fair degree of accuracy, the nature of the bottom sediment merely by observing the depth at which the sounding line stopped paying out. And it was on this first Atlantic transect that they started to formalize their notions about the nature of the ocean floor. They noted that Pteropod ooze seemed to predominate at the shallowest depths, down to about 400 fathoms; that the white Globigerina ooze extended down to a depth of about 1,500 fathoms, where the transition to the gray ooze began to occur; and that the gray ooze in its turn gave way to the zone of red clay at a depth of about 2,200 fathoms.

Because of the strict depth dependence of these zones they also noted that they could, rather easily, be related to the undersea geography of the Atlantic. The first four stations west of Tenerife— some 300 nautical miles—showed classic white Globigerina ooze. The next two stations, out to about 750 miles from Tenerife, showed the gray ooze. Beyond that, out to the start of the rise to the mid-Atlantic ridge, was the red clay. The ridge itself was covered in Globigerina ooze. To the west of the ridge, the situation was a mirror image of that to the east, with the ooze of the ridge giving way to gray ooze that in turn gave way to red clay. As they approached the

West Indies and depths decreased, the white Globigerina ooze gradually returned. Wyville Thomson wrote in 1874, "The general consensus of so many observations would go far to prove, what seems now to stand, indeed, in the position of an ascertained fact, that wherever the depth increases from about 2,200 to 2,600 fathoms, the modern Chalk formation of the Atlantic and of other oceans passes into clay." The Scientifics had discovered one of the most fundamental properties of the ocean, the calcite compensation depth.

The calcite compensation depth is the phenomenon that is responsible for the geographic and depth distribution of the white Globigerina ooze that Wyville Thomson and the other Scientifics discovered on that first fateful transect across the Atlantic. Put simply, the acidity of the ocean increases with depth and there is a level in the ocean where the water becomes sufficiently acidic to dissolve away calcium carbonate, the primary constituent of the animal shells that make up the Pteropod and Globigerina oozes. In technical parlance, the calcite compensation depth is the particular depth in the ocean where the rate of calcium carbonate supply from the surface is balanced by the rate of dissolution so that there is no net accumulation of carbonate. Imagine the undersea mountains of the world's oceans having the equivalent of a snow line and you have an excellent mental image of what Wyville Thomson and the rest of the Scientifics had discovered. Indeed, as Figure 5 shows, they had inadvertently picked just about the best place in the world's oceans to see the effect of the calcite compensation depth, because the changing nature of the bottom sediment could so easily be traced out to the mid–Atlantic ridge and from there to the American continental shelf.

For all the obviousness of the changing nature of the bottom sediment, Wyville Thomson and John Murray had no clear idea what it was that caused the sediment transition with increasing depth. Yet Wyville Thomson was presciently close to the truth when he described an experiment that Buchanan had done to prove that

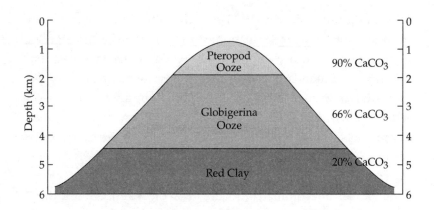

FIGURE 5 Distribution of calcareous sediments

the presence of the red clay was caused by the *lack* of ooze rather than the *addition* of red clay from the continents. Buchanan had poured weak acid on a sample of white Globigerina ooze and noted within minutes "that there remained, after the carbonate of lime had been removed, about one per cent of a reddish mud consisting of silica, alumina, and the red oxide of iron." From this Wyville Thomson was able to conclude ". . . that the red clay is not an additional substance introduced from without . . . but that it is produced by the removal, by some means or other . . . of the carbonate of lime which forms probably about 98 percent of the material of the Globigerina ooze."

Today we know that the calcite compensation depth is a complex threshold in the ocean whose depth varies under precise conditions that are regulated by the acidity of the seawater. At its simplest, the CCD (as it is customarily abbreviated) is at shallowest in the northern North Atlantic and at its deepest in the northern North Pacific. One of the main regions of deepwater formation today is the Greenland Sea. Close to the area of their formation these waters are well oxygenated, cold, and supersaturated with dissolved calcium carbonate. As they start their long journey away from the area of their formation, organic carbon from dying and decaying surface algae is progressively added to these deep waters in the form of mildly acidic carbon dioxide. This makes deep waters more and more acidic the further they are from the area where they were formed and, therefore, progressively more inimical to the deposition of calcium carbonate. The result of this is that the depth of the CCD between the Atlantic and the Pacific is an inclined plane with the deepest part in the northern Atlantic at a depth of about 5.5 kilometers and the shallowest part in the northern Pacific at a depth of only 1 kilometer.

The CCD is one of the most important and fundamental properties of the deep ocean and, as with so many other marine phenomena, we owe our knowledge of it to the voyage of *Challenger*.

ARRIVAL IN THE NEW WORLD

St Thomas, Virgin Islands, Danish West Indies, March 14-16, 1873, 18° 22′ N, 64° 56′ W

It was squalling heavily from the southeast when *Challenger* sighted the small island of St. Thomas on March 14, 1873. Rain drove in great sheets across the bow as the ship struggled to drop anchor in the outer harbor. The storm had raged for several days, and the crew was thrown about mercilessly. Only those on watch braved the upper deck. Everybody else stayed in bed. Of those, only the bluejackets and junior officers like Herbert Swire were halfway comfortable in hammocks, which could roll with the tossing of the ship, and Swire pitied those of higher ranks with the dubious privilege of a fixed berth. Then, quite suddenly, the storm vanished and a thick fog formed. Those crewmen who were tardy coming up from below decks found that the first sight of land for four weeks had been suddenly obscured.

The bluejackets and officers knew that the crossing should have been brisk with the easy trade winds they'd picked up as soon as they left Santa Cruz, but the pestilential Scientifics had insisted on their stops every 200 miles for sounding and dredging. It was enough—more than enough—to irk a seaman's soul and the sharks that had followed them for the last 1,100 miles of their journey were an unwelcome portent. One, a 20-foot tiger shark made the bluejackets very nervous. To them he seemed too clever by half, swimming up to the lump of salt pork that the Scientifics dangled overboard, sniffing and nudging it disdainfully before swimming away. But let anything fall overboard and John—as Joe Matkin had named the shark—was there "like a lawyer" devouring the morsel in great, slicing bites. There were those aboard who said that John's untiring presence was a sure sign that somebody would go overboard before they made the West Indies.

And then there was the attitude of the Scientifics. Since leaving Portsmouth they'd kept themselves to themselves, but now that they were getting their specimens on a regular basis they had insulated themselves to the point of standoffishness. All on board felt it to a greater or lesser degree. They would like to know a little more about the reasons for their sacrifice—three and half years away from hearth and home after all! The feeling became so intense that in the first week of March the captain himself prevailed upon the Scientifics' leader, Professor Wyville Thomson, to discourse on the findings of the expedition so far. The lecture was, all agreed, most interesting and the professor promised to make these talks a regular feature. Joe Matkin even went so far as to transcribe the lecture word for word and send it home to his family in Rutland, but he was known as a grammar-school lad and something of a swot.

That was not the only incident on the voyage. The thief was still up to his old tricks and not many days previously had broken open the officers' meat locker in the dead of night despite the presence of a heavily armed marine outside the cabin. How he got in was a mystery and the officers were in a rage. And that, too, was a symptom of a continuing source of tension: The officers were eating as if they were still in harbor while the bluejackets were dining only on salt meat and biscuit. Only the daily allowance of lime juice and grog offered anything palatable. They were, as Matkin put it, "fast coming down to our precise fighting weight."

Captain Nares elected not to get too close to the township of Charlotte Amalie, the island's capital, and anchored in the outer harbor known as Gregorie Channel. Having been to St. Thomas before, he well knew its reputation as a virulent source of yellow jack (yellow fever). The Danes, the island's colonial governors, had already taken steps, in their efficient Scandinavian manner, to eradicate the disease by widening and deepening the channel between the inner and the outer harbor. Now the tidal swell emptied the inner bore of the sewage that collected there daily, washing it out to sea and greatly improving the amenity of the island. But still Nares would not take the risk. Several of the Scientifics had taken out private health insurance and these, without exception, contained exemptions—ports whose level of public health was so poor that the insurance company would not underwrite the risk.

As he stared down from the narrow bridge at the bustle below, Nares reflected that nothing would damage his career faster than losing one of the scientific staff in his charge. And there were, he knew, other ports far worse than this one. Indeed, any of the ports along the "White Man's Grave"—the bluejackets' name for the west coast of Africa—were a far worse bet for a sailor's continued good health than this place. So on the way to Africa he planned to avoid all but Cape Town and Simonstown. But here in Charlotte Amalie he would take no chances either and so, as they waited for the fog to lift, they swung gently at anchor in the outer harbor.

For most of the crew, the view—when it came—was worth waiting for. The little township of Charlotte Amalie on the waterfront of the inner harbor was backed by tall thickly wooded hills. For William Spry, confined as he was for so much of his time to his engine room, it was a relief to see "a variety of trees, all green and tempting, as far as the eye could reach" after so many weeks at sea.

Beyond the hills, and dominating the center of the island, was a curious saddle-shaped peak fully 1,500 feet high.

On its flanks, and penetrating in some places almost all the way down to the shore, were vast fields of scrubby undergrowth, the remnants of the sugar-cane plantations that been the island's staple industry until 1848 when the industry collapsed because of the abolition of slavery. All but one of the island's plantation owners had been bankrupted by the emancipation, receiving only £50 for each freed slave from the government. Both Swire and Matkin noted that the transition to freedom had not done the island's economy much good; the Negroes worked only a bare minimum, preferring odd jobs, especially coaling ships, to regular work. St. Thomas had become an important port of call in the West Indies, because it served as the sorting office for the considerable amount of mail that arrived via a variety of shipping lines each month. Yet, both coaling and the mail were vast economic improvements on the island's more distant past. Two hundred years earlier, St. Thomas was notorious as the focal point for the pirates of the Spanish Main and, more recently during the Napoleonic wars, was a rendezvous for British merchant ships waiting to convoy across the Atlantic on the long and dangerous route to Britain.

Swire loved St. Thomas, despite its checkered history, because the waters around the island were dotted with dozens of smaller islands, some only a few feet across, that the locals called cays. Their outer rims were formed by a profusion of corals that had grown up to the water surface and these, in turn, supported a wide variety of other lime-secreting animals, such as mollusks, that raised the cay above the level of the surrounding sea. This mass of coral and invertebrates was then bound together by the limey secretions of calcareous algae, making a tiny landmass that could withstand the ravages of the local hurricanes and where the footing was easily firm enough for exploration. The lagoon formed in the center of this circle would eventually silt up, giving the cay a firm flat interior that was perfect for sunbathing and picnicking. The waters around these cays themselves contained an abundance of submerged corals in the form of sheets and branched horns and these, too, lived in communion with tough strands of lime-secreting algae, invertebrate shells, rock-mantling bryozoans, and a confusion of brightly colored fish to gladden the soul of even the gruffest tar aboard.

HMS *Challenger* had arrived in the kingdom of lime.

THE SYMBIOTIC ECOSYSTEM

The kingdom of lime is the kingdom of the reef, very ancient ecosystems that are found in many places in the tropical oceans of the world. Today, coral reefs occur in the Americas, particularly the Caribbean; in the Red Sea, the Persian Gulf, and around the Maldives in the Indian Ocean; further to the southwest around the Seychelles, Mauritius and off the east coast of Africa; and in east Asia and the tropical Pacific. But these reefs are not all the same; there are significant regional differences in the number and type of species that make up the Pacific reefs and those in the Atlantic and Indian Ocean. Western Pacific colonies tend to be much more diverse than their counterparts, with up to 75 percent more genera and 85 percent more species.

Modern reefs, whether in the Pacific or elsewhere, are among the most diverse ecosystems on our planet. Not for nothing have they been called the "rain-forests of the sea." And yet, what exactly are reefs? To most of us, familiar as we are with a thousand Jacques Cousteau and David Attenborough TV documentaries, a reef is a familiar feature of the ocean. Mention it and instantly images of fine white coral sands, swaying palm trees, Fletcher Christian, and South Sea Islands are conjured up, images not dissimilar to those about which Herbert Swire wrote. Today, reefs occur mainly in the shallow waters of the tropics, where there is abundant sunlight to fuel the tiny symbiotic algae that live inside the tissues of the coral animal, the polyp, and that provide it with nutrients through photosynthesis. But not all reefs today, or indeed many known from the fossil record, are photosymbiotic.

Non-symbiotic corals have been an important constituent of reefs since at least the beginning of the Phanerozoic eon. During the Archaean and Proterozoic (the time before the evolution of animals with hard parts), there were vast agglomerations of reefs that were composed, not of corals, but of bacteria-like organisms known as blue-green algae.

An important concept common to all reef ecosystems, whether in the present or the past, is the framework builder. Framework builders are the component of the reef ecosystem that produce most of the hard lime or calcium carbonate that gives a reef its structure. In the Proterozoic, the framework builders were the blue-green algae, while in the Phanerozoic (the era since the evolution of animals with hard parts 550 million years ago) they have been any one of a variety of invertebrate species. In fact, one of the things that makes the history of reefs such a complex and confusing subject is the fact that the main framework builder has changed regularly across the course of geological time and superimposed on top of these changes are smaller changes in the composition of the animals and plants (the predators and prey that make the coral world their home) that co-habit with the framework builders.

These variations have led to a lot of confusion over the seemingly simple question as to what, exactly, a reef is. Part of this confusion comes from the way the term is used by geologists (as distinct from biologists). To many geologists, a reef is simply a thick-ened mass of limestone that is quite distinct from the underlying and overlying rock; and to a further subset of geologists, a reef is synonymous with the concept of oil, for fossil reefs are often associ-ated with hydrocarbons. This notion is especially prevalent in the North American oil exploration community, where the bottom line has always been, well, the bottom line. To an oil geologist, a reef is merely a potential fuel tank, and its biological antecedents are, at best, completely irrelevant. So for many geologists, particularly of the older generation, the term "reef" is more descriptive than any-thing implying a biological connotation. But that was not the understanding that prevailed aboard *Challenger*, because the Scientifics carried with them the recent and hard-won knowledge of the greatest biologist of them all. Charles Darwin, in the days before he wrote *Origin of Species*, used the observations that he had made on his journey around the world aboard *Beagle* to develop a theory of how reefs, particularly those that form coral atolls, grow. He showed that reefs in the Pacific grow atop volcanoes that are subsiding below the surface of the ocean. As the volcano sinks, the growing reef first forms a fringe around the island and finally, as the volcano sinks beneath the surface, all that is left is a more or less circular reef. Classic examples of such islands, scattered across the Pacific, are Majuro and Kwajalein in the Marshall Islands. Kwajalein's ring-shaped lagoon is used as a target for intercontinental ballistic missiles launched from Vandenberg Air Force Base in California.

It was Darwin's theory that proved the turning point and for biologists since Darwin, the reef concept has been different from that of geologists. To biologists today, a living reef is a colony of tiny animals, polyps, which comprise corals, which in turn live in conjunction with a host of other organisms, together making up the reef ecosystem. Corals and many other reef-dwelling organisms secrete a hard skin of calcium carbonate that makes them the framework builders of the reef.

Coral polyps live symbiotically with algae called zooanthellae. The algae derive nutrients from the animals in the form of nitrogenous waste and carbon dioxide produced by the animals' respiration, while the coral animals also impart protection and a safe haven in which the algal cells can comfortably make sugars by photosynthesis. In return the corals receive and use the products of photosynthesis; sugar and oxygen.

Other framework builders that contribute to the reef ecosystem include bryozoans and certain types of algae. Together, these build a microcosm in which mollusks, starfish, sharks, eels, and fish thrive in an abundance of color and warmth that collectively make up one

of the most beautiful ecosystems on Earth. It might be the relationship between the zooanthellae and the coral that is responsible for
the rather restricted distribution of shallow-water reefs to 30 degrees north and south of the equator and the western sides of the
ocean basins. In these regions, the ocean circulatory system keeps
the nutrient content of surface waters low so that free-floating algae cannot flourish and the waters remain relatively clear and transparent to sunlight. This creates ideal conditions for the symbiotic
relationship between polyps and zooanthellae.

Coral polyps are members of a strange group of animals known
as the Cnidaria, pronounced "ni-daria." The cnidarians have only two
layers of tissue; the rest of the animal kingdom (with the exception
of the sponges) has three. The two layers of tissue are known
formally as the epidermis and the gastrodermis. The names are give-
aways: The epidermis is the external skin that protects the animals,
the gastrodermis is the internal layer that digests food. Between the
two tissue layers of the cnidarians is a layer of jelly, the mesoglea,
which provides mechanical support for the rest of the body.

The two-layer (technically known as diploblastic) body plan
tells us that the cnidarians, and the corals, are a very ancient group
of animals. We can think of them as an experiment in evolution that
was later superseded by a better design, the three-layered (or triplo-
blastic) body design on which all higher invertebrate animals and
all the vertebrates are based. But the diploblastic design cannot be
all that bad for the reef-building corals, after a long and illustrious
history, are still with us today.

However, it seems likely that photosymbiont-bearing corals
evolved only 65 million years ago. Apart from their requirement for
clear waters with sufficient sunlight for the symbionts to flourish,
photosymbiotic corals, like their deeper-water cousins, must also
inhabit water supersaturated with dissolved carbonate to enable the
coral polyps to form the protective limey skeleton that is the basis
of the reef ecosystem.

The West Indies was not the last place that *Challenger* encountered reefs on its long journey. Their next port of call, Bermuda, was famous for them, and within 18 months, off northeastern Australia, *Challenger* would explore the most spectacular coral reef system in the world.

St Thomas, Virgin Islands, Danish West Indies, March 16 , 1873, 18° 22' N, 64° 56' W to Hamilton, Bermuda, Atlantic Ocean, May 9, 1873, 32° 18' N, 64° 48' W

After completing their examination of the coral cays of the Danish West Indies *Challenger* took on coal, food, and water at St. Thomas while the crew contemplated the next leg of their mission, the voyage north to Bermuda and then beyond to the east coast of North America. Just as they were about to set sail, however, intelligence reached them via dispatches that an English vessel was in their area and in need of assistance. On March 23, 1873 *Challenger* hauled in the anchor and set off in search of the ship. They soon found her, *Varuna*, out of Liverpool. She had departed New York the previous January but had run into stormy weather and lost her main and mizzenmasts before the crew took to the lifeboats. At some point after that she was boarded by a prize crew who sailed her to within 15 miles of St. Thomas, stripping her of everything that was not nailed down before abandoning her to her fate again. Taking her in tow, *Challenger* triumphantly sailed her into the little harbor at Charlotte Amalie where she was ceremoniously handed over to the British Consulate.

The next day, March 24, the crew were under orders of science again and set sail on the next leg of their voyage. It was here that disaster struck and the *Challenger* expedition sustained another fatality. As they were pulling in the dredge, the rope snapped. The tension on the rope, coupled with that of the massive india-rubber accumulator that protected the mast, was so severe that "it carried away an iron block that was screwed in to the Deck," wrote Joe Matkin, "it struck a sailor boy named William Stokes on the head, and dashed him to the deck with such a terrible force, that his thigh was broken, and his spine dreadfully injured." Stokes died of his injuries a few hours later. When the dredge was finally hauled in, it was found to contain a few sad scraps of coral as well as the ubiquitous Globigerina ooze, poor compensation for the loss of so young a life. Two days later, after evening quarters, the crew assembled to say their last goodbyes to the young seaman. The bell tolled softly as the body, wrapped in a lead-weighted hammock as was the custom, slid gently overboard "to be seen no more until that day when the sea shall give up its dead," as William Spry sadly put it.

The death was a shock to all on board and it was a subdued company that made sail for Bermuda. But the spirits of the bluejackets were raised on April 1 when "Hands to Bathe" was piped and 80 of them went over the side to swim, protected by a gig on shark-watching detail. A fine time was had by all. Other entertainment was provided by the flying fish that chased the ship. Joe Matkin wrote, "On a dark night, if a lighted candle is placed in the ship's port they will often fly in."

By March 26, 1873, *Challenger* was 85 miles north of St. Thomas and sailing north along the easternmost boundary of an area that within a hundred years would strike terror into the hearts of seamen and aviators alike: the Bermuda Triangle.

Climate Triggers and Bermudan Secrets

St Thomas, Danish West Indies, March 16, 1873, 18° 22′ N, 64° 56′ W
to Hamilton, Bermuda, 32° 18′ N, 64° 48′ W Atlantic Ocean

THE CLIMATE BOMB

The Bermuda Triangle is, of course, famous for the unexplained ship
and plane disappearances that have occurred there. Even today you
will not find it on any map. Broadly speaking, the Bermuda Triangle
is defined by three points: the island of Bermuda, the city of Miami,
and the island of Puerto Rico. In March of 1873 *Challenger* was
traveling north along the right-hand side of that triangle.

It is difficult to determine the truth or otherwise of the
Bermuda Triangle myth. It's tempting to regard the problem as
simply statistical. In the days when radio and radar communications
were primitive or nonexistent, planes and ships were lost at sea all
the time; it was one of the perils of the job. But the enduring myth
of the Bermuda Triangle as a region where this happens rather too
often for comfort will not go away. Many explanations, which range
from the feasible to the plain daft, have been advanced. An example
of the former is that the Bermuda Triangle lies on the 80° meridian,
a line that is one of two places on Earth where the magnetic and
true north poles are in perfect alignment. Navigators on these so-
called agonic lines do not need to make the usual correction to

their compasses. It is just possible that even experienced navigators might be thrown if they failed to re-adjust their compasses when on an agonic line. The other agonic line, in the Pacific Ocean just to the east of Japan, is also known to local mariners as a place of unexplained disappearances.

The Bermuda Triangle is also known as a region where unpredictable weather can develop rapidly. It is, after all, in the region where the Gulf Stream originates and that vast mass of water moving northeast at 4 knots carries a hefty kinetic punch. Small "meso" thunderstorms of ferocious intensity can develop and vanish quite suddenly in the Caribbean.

At the daft end of the scale of Bermuda Triangle explanations is the notion that the area is frequented by roving extraterrestrials who abduct humans and their machines. Indeed Steven Spielberg's epic, *Close Encounters of the Third Kind*, opens with the famous loss of Flight 19, when five Grumman torpedo bombers disappeared without trace in 1945, and adds the fantasy of them turning up in the Mexican desert 30 years later.

Yet there is another explanation for the mystery of the Bermuda Triangle. This is the conjecture that the ocean—and indeed the air above it—can suddenly lose the ability to support objects such as ships and aircraft because the density of water and air are suddenly reduced. This idea proposes that the density change is caused by the submarine convulsions of a strange and little-understood material buried underneath the seabed: methane hydrate.

Methane hydrates are part of a group known as the gas hydrates that occur when a gas molecule is surrounded by a "cage" of frozen water molecules. Technically they are clathrates (crystalline solids), similar to ice, except that they are not of uniform molecular com-position: Part of their structure is provided by a "guest" molecule. In the case of methane hydrates the guest is a molecule of one of the most flammable naturally occurring substances: methane. The name clathrate comes from the Latin *clathratus*, which means "enclosed by bars or grating." They have been known since the early nineteenth century. Indeed it was Sir Humphrey Davy who first synthesized them, in the form known as chlorine hydrate, in his laboratory, from a chilled mixture of chlorine gas and water. In 1823 Michael Faraday published a paper entitled "On Fluid Chlorine" in the *Philosophical Transactions of the Royal Society*, a journal to which Wyville Thomson regularly contributed during the *Challenger* expedition.

Methane hydrate, the most common of the naturally occurring hydrates, comprise a truly vast energy reservoir. It is estimated that existing deposits contain enough energy to fuel the world for the next 350 to 3,500 years. The huge uncertainty of this estimate, though, reflects just how little we know about their distribution. They occur in two principal settings, on land in the high latitudes of the Arctic and the Antarctic, where the intense cold keeps the molecule stable in the permafrost, and in the deep ocean, on the outer continental shelves where the pressure of the overlying water keeps the clathrate stable. These two environments give us an important clue to the properties of methane hydrates: they can exist, or to be more accurate, remain stable, only within a narrow range of pressure and temperature. If the pressure is reduced by a few tenths of a bar or the temperature increased by only a couple of degrees Celsius, the hydrate spontaneously decomposes, releasing vast quantities—more than 160 times their crystalline volume—of flammable methane.

Until recently methane hydrates were regarded as little more than a laboratory curiosity or a nuisance when they caused blockages in gas pipelines. That changed in 1964 when a Russian drilling crew working in the northern Siberian gas field of Messoyakha discovered naturally occurring methane hydrates. Further prospecting revealed vast quantities under the northern Siberian Tundra, but it was not until the 1970s that methane hydrates were found in their other natural habitat: the oceans, where they were discovered by the Deep Sea Drilling Project's vessel, and HMS *Challenger's* namesake, the *GLOMAR Challenger*.

The ship was drilling off the coast of Guatemala when it unexpectedly penetrated a methane hydrate deposit. Previous cruises had logged the presence of vast seismic reflectors below the seafloor that were so dense that they were often mistaken for the seafloor itself. But these bottom simulating reflectors (BSRs) turned out to be the lower limits of undersea methane hydrate deposits. The extent of these BSRs is a clear indication of just how much methane hydrate is under the ocean floor.

BSRs exist because methane hydrate deposits have a relatively high seismic velocity; sound waves speed up when they encounter a methane hydrate reservoir. An echogram of the kind pioneered by Maurice Ewing shows this clearly. This velocity difference distinguishes hydrates from overlying sediment as well as the underlying BSRs themselves, which often cap reservoirs of natural gas. Before the GLOMAR Challenger's cruise, smaller ships had penetrated these BSRs by coring, but they had not retrieved any methane hydrate because, relieved of its imprisoning pressure and temperature, the methane hydrates spontaneously decomposed on the way to the surface, vanishing into thin air before the researchers could even get a look at them. But the GLOMAR Challenger successfully retrieved a 3-foot length of core that contained methane hydrate and shipped it to the Colorado School of Mines, a leading center for energy studies in the United States, where it sparked a storm of interest.

Where do the methane hydrates occur around the world? Mapping their distribution is a complex high-tech task and it is no surprise to learn that their distribution around North America is particularly well known. One of the richest fields discovered so far is on the northern margin of the Bermuda Triangle on the seafloor topographic high known as the Blake Plateau.

With methane hydrate fields being discovered all the time, interest in them as a new form of energy is snowballing. A special leg of the Ocean Drilling Program (the successor to the Deep Sea Drilling Project) was recently commissioned to investigate the methane hydrates under the Blake Plateau. One of the most interesting discoveries made by the scientists who studied the Blake Plateau material was the age of the methane hydrates. Using the radioactive decay of iodine they were able to determine their age as 55 million years old. This age is notable because it is precisely the age of the boundary between the Paleocene and Eocene epochs. For many years the P–E boundary (as it is commonly abbreviated) was not considered an important epoch boundary. But in the late 1980s Leg 113 of the Ocean Drilling Program drilled the Maud Rise near Antarctica and discovered a major change in the proportion of oxygen isotopes in sediments that spanned the P–E boundary. Put simply, a carbonate shell (such as a foram) growing in warm water incorporates more of the light isotope of oxygen (oxygen-16) than a foram shell growing in colder water, which incorporates relatively more of the heavy isotope (oxygen-18). Since the 1950s this approach has been developed into a fantastically sensitive paleo-thermometer that can now recognize temperature differences of less than 1°C.

Lowell Stott and James Kennett, working at the University of California at Santa Barbara, were stunned to find that the temperature record for the period across the P–E boundary in the deeper core was *warmer* than the record in the shallower core. This was quite the opposite of what they expected, because the temperature of the oceans generally decreases sharply with depth. How could deeper waters be warmer than shallower waters? Their explanation was that during this one critical interval of the Cenozoic (the last 65 million years of Earth history) deep waters were not forming in the high latitudes (as is the case today) but rather in the low latitudes and flowing away from the tropics, eventually arriving in the Antarctic and warming the deep waters there.

In the present day, as in much of the Cenozoic, deep waters are formed near the poles by surface waters that cool and sink as their density increases. These then flow out from the high latitudes, circulating cool water throughout the world's ocean basins. What could have caused deep waters to form in the low latitudes and disrupt the usual system of deepwater circulation so drastically? Stott and Kennett suggested that the answer was massive global warming, so intense that it was able to increase the density of low-latitude waters through enhanced evaporation, which concentrated it into something approaching brine. But this merely pushed the question back, for what could have caused such massive global warming? This time the answer lay in the carbon isotope record in the core. Stott and Kennett discovered that at exactly the same part of the core where the oxygen isotope anomaly had been found there was a large negative anomaly in carbon isotope values. There are two stable isotopes of carbon, carbon-12 and carbon-13, which differ only in the number of neutrons in their atomic nuclei. The former has 6, the latter 7. Conventionally the proportions of the two are expressed as a ratio normalized against a known standard. Where there is more of the light isotope of carbon the ratio is more negative, where there is more of the heavy isotope of carbon the ratio is more positive.

Stott and Kennett discovered that the normal ratio of carbon isotopes in the natural world was quite suddenly swamped by a preponderance of the lighter of the two isotopes—carbon-12. Normally the proportion of the two isotopes of carbon is controlled by the rate of photosynthesis in plant tissues. When photosynthesis proceeds more rapidly, proportionally more of the light isotope of carbon, carbon-12, is squirreled away into protoplasm. When this occurs in the ocean, the surrounding water is enriched in the heavy isotope of carbon and this is reflected in a higher ratio of carbon-13 in the hard parts of shell-secreting animals such as forams. But the carbon isotope ratio in the shells of the forams that Stott and Kennett analyzed was vastly more negative than could reasonably be explained even if there had been a total collapse in the rate of photosynthesis. Also, the shift was found to be equally large in forams that came from deep waters, which proved conclusively that the light carbon must have come from a reservoir outside the pool normally used by photosynthesis.

Methane has a carbon isotope composition about 60 times more negative than the isotopic composition of carbon in the ocean and the biggest reservoir of methane in the world is methane hydrate. If that methane were suddenly liberated from the hydrate form, the impact on the carbon isotope record would be immediate and severe. Indeed, modeling experiments have shown that it would be exactly the same as the record that was found in the deep-sea cores from the Maud Rise.

At about the same time that a young American geologist named Jerry Dickens was developing these insights, more ocean drilling was bringing up other material of P-E boundary age, from the mid–Pacific as well as the Blake Plateau area itself. All three records showed about the same magnitude of carbon isotope shift at the P-E boundary. At the same time, work on land led by the Swedish geologist Birger Schmitz showed an anomaly in land sections of P-E age around the Mediterranean region. The evidence was unequivocal: There had been explosive outgassing of methane hydrate deposits at the P-E boundary, enough to cause a global warming of up to 8°C. As time went by, the details of the event were elaborated. Santo Bains and I, in our Oxford Laboratory, in conjunction with Richard Norris of the Woods Hole Oceano-graphic Institution, showed that there was not one event but several—a cascade of methane hydrate outgassing events, probably with each one triggering the next. We found, however, that the initial triggering event was nothing more sinister than an under-water earthquake, the kind of thing that happens all the time, especially close to the mid–ocean ridges.

But what had stopped this runaway greenhouse event once it started? Again it was detailed work led by Santo Bains that provided the answer: Bains analyzed the concentration of the element barium in sediments deposited across the boundary and showed that it increased markedly and then dropped back down again in perfect synchrony with the carbon and oxygen isotope records. The expla-nation was devastatingly simple. Vast amounts of methane had been pumped into the atmosphere, causing greenhouse warming. But methane rapidly decomposes into carbon dioxide, which was then consumed by this vastly increased mass of photosynthesizing tissue greedily sucking all the liberated carbon dioxide back down into the ocean again.

What is the significance of the age of the Blake Plateau methane hydrates? They could not have spontaneously dissociated to plunge the world into the closest analogue we know to our own greenhouse future. Methane hydrates take eons to form, as dead organic material is cooked and compressed into raw methane gas and then locked up in a cage of enclosing water-ice crystals. The hydrates that let go at the P-E boundary must have predated it by millions of years. Is, then, the fact that the Blake Plateau hydrates are precisely the age of the P-E boundary a coincidence? No, it seems likely that the methane hydrates of the Blake Plateau today are themselves the record of the CO_2 that was released at the P-E boundary and that was eventually drawn down by the enhanced photosynthesis suggested by Bains and co-workers. The clue to the day the oceans boiled is buried under the Bermuda Triangle, and who can say that these methane hydrates are not too awaiting the day when they will once again be unleashed?

Since the hypothesis was first put forward in the early 1990s, methane hydrates have been implicated in mass extinctions elsewhere in the fossil record. For example, at the Cretaceous-Tertiary boundary (more usually associated with a probable meteor impact), 65 million years ago when the dinosaurs died out, the sudden climate change at the Cenomanian-Turonian boundary (90 million years ago), the climatic oscillations of the most recent series of ice-ages (2.5 million years ago to the present day) as well as the biggest series of glaciations of all time: those that characterized the late Neoproterozoic eon (600 million years ago).

Methane hydrates are not the only clathrates that can form in the deep ocean environment. Carbon dioxide hydrates can be induced to form in deep waters and the current thinking is that, if artificially induced, these deposits might be able to serve as a storage medium for humanity's excess CO_2. Peter Brewer of the Monterey Bay Aquarium Research Institute and his associates have successfully induced CO_2-hydrate formation at depths greater than 3 kilometers. At such depths, CO_2 hydrates are metastable, that is, that as long as the pressure and temperature remain unaltered they remain in this state forever. Brewer reasoned that if CO_2 from power plants were piped directly to the deep sea at the appropriate water depth, it could be stored there indefinitely in the form of hydrate so that it would not contribute to global warming.

Of course this is still experimental and much work must yet be done to determine this plan's feasibility. But even if the idea is sound, there remains the question of whether we should do it. If there is one thing that we have learned from the legacy of the seafloor's fossil record—and particularly the sediments that straddle the P–E boundary—it is that hydrates are not necessarily stable. They are subject to periodic earthquakes and volcanic activity that can destabilize them. It is likely that their eventual dissociation is simply a matter of time and statistics. Carbon dioxide hydrates stored in reservoirs on the seafloor would probably require protection as strong as that of a nuclear waste reprocessing facility to prevent an accident that would plunge the world into a climatic crisis dwarfing the one that took place at the P–E boundary.

Finally, returning to methane hydrates, there is no question that their energy potential is enormous. With confidence in nuclear power plummeting and environmental concerns about conventional oil and gas extraction growing daily, methane hydrates offer a comparatively cheap and easy form of new energy. But how is it to be extracted? Some think that the approach will be similar to that used to recover petroleum from difficult sites: Steam or hot water could be pumped down a drill hole to melt the hydrate, which would be collected from another drill hole. The resulting methane gas could then be piped ashore.

The problem with this method is that dissociating methane out at sea could easily destabilize the continental shelf that supports the rig and, as we have seen, the larger scale consequences of a methane hydrate "event cascade" could be catastrophic. This hazard has led some scientists to suggest that retrieving the hydrate intact and then liquefying it on ships or drilling platforms is a better approach. One suggestion is to burn the hydrate to form hydrogen and carbon monoxide and then use a catalyst to convert the mixture into a liquid hydrocarbon, which could be readily transported by ship. The downside, though, is a 35 percent loss of energy.

Another approach is being considered by Roger Sassen of Texas A&M University, who envisions production on the ocean floor. Extracted methane could be recombined with water to form a new hydrate uncontaminated by mud and rock. Submersibles would then tow the hydrate in special storage tanks to shallower areas where it could be more safely decomposed into water and fuel.

Whatever the eventual extraction method, it is clear that methane hydrates are the fuel of the future. A harder question to answer is whether methane hydrates are responsible for the myth of the Bermuda Triangle.

Hamilton, Bermuda, April 4, 1873, 32° 18′ N, 64° 48′ W

It is one of the ironies of the *Challenger* expedition that despite the many phenomena and organisms that it did discover—the calcite compensation depth, the mid-Atlantic Ridge, manganese nodules, to name but a few— there were also spectacular secrets of the silent landscape that it did not uncover and one of these was the existence of methane hydrates. So, unaware of what they had missed, the voyagers arrived in Bermuda on the evening of April 4, 1873, sliding to anchor at Grassy Bay with the aid of a local pilot standing at the foretop and directing the four men at the wheel. The narrows through the reefs were treacherous and a knowledgeable pilot was a requirement. There were dozens of barely submerged corals all around them and Joe Matkin could see on many of them the impaled remains of less fortunate ships.

The island of Bermuda was strategically important to Victoria's navy and the whole of the British North American Fleet was stationed there under the command of Governor Major General Lefroy. Because Bermuda was so important, the main town of Hamilton had the most elaborate docking facilities in the North Atlantic. These facilities included an enormous floating iron dock that had been recently towed all the way from Britain by four men o' war, among them HMS *Warrior*. Built in 1861, *Warrior* was the jewel in the British navy's crown. She was the ultimate expression of British naval supremacy, more than 400 feet long and armed with 26 muzzle-loading 68 pounders and 10 breech-loading 110 pounders. With a hull composed of iron plate she was almost single handedly responsible for keeping the peace with the French throughout the 1860s despite much saber rattling. But by the time *Challenger* arrived in Bermuda in 1873, *Warrior* had already been decommissioned from front line service. She had been rendered obsolete by the development of the fully steam powered *turret* warship based on the fearsome *Monitor* design pioneered by the Americans during their own civil war less than a decade before.

The harbor at Grassy Bay was only eight miles from the Bermudan capital, Hamilton, and all on board found the location convivial. William Spry wrote, "Nothing could have been more romantic than the little harbor stretched out before us: the variety and beauty of the islets scattered about; the clearness of the water; the quantity of boats and small vessels cruising between the islands, sailing from one cedar grove to another, made up as charming a picture as could well be imagined." It was a welcome relief to the tedium of dredging and nothing short of blissful to have something to take their minds off the death of young Stokes only 10 days before. It was almost unbelievable, therefore, that amidst such tranquility and beauty, death should once again visit *Challenger* on their first night in port.

This time the victim was one whose continued well-being was central to the comfort and education of the youngest member of the ship's company. Joe Matkin had just retired for the night when he was woken by the ship's writer Richard Wyatt in a state of great agitation. In the quiet of the crew's mess, among the snores and grunts of a score of sleeping men, Wyatt heard strangled gasps coming from the hammock of ship's schoolmaster Adam Ebbels, the man charged with the education of Captain Nares's young son until they reached Australia. When Wyatt went to investigate he found the schoolmaster lying in his hammock, hands clawed, ashen-faced, and quite, quite dead. An inquest held by the ship's surgeon in the morning found that he had died of apoplexy, or what we today would call a stroke. He was buried in a small nearby cemetery less than seven hours after his death.

There was more bad news. The mail brought news that the steamer *Atlantic*, one of the original five vessels of the White Star Line (the shipping line whose name would become forever associated with marine tragedy when *Titanic* sank several decades later) had sunk near Halifax on April 1st, running aground during one of the storms that plagued the region, with the loss of 560 passengers. The news triggered a rush of morbid sympathy below decks. Then, as now, there was a shared sense of peril among those who made their living at sea. This tragedy, coupled with the recent deaths of William Stokes and Adam Ebbels, was a potent reminder to all aboard of the dangers inherent in their enterprise. For Joe Matkin that feeling was compounded after Sunday church when he walked in the dockyard graveyard. He was struck by the number of naval men buried there. "You would be surprised to see what a quantity of seamen and Naval Officers have been buried there during the last 80 years," he wrote in a letter to his cousin Tom. "There are 80 or 90 different Ship's gravestones; each ship has a large stone with the names of all the officers, seamen and marines, that they have buried engraved thereon. Drowning, falling from Aloft, and Yellow Fever appear to have caused the most deaths. . . ." It was another reminder that a three-and-a-half-year voyage round the world, even in the year 1873, was no picnic.

THE RIVER OF HEAT

Bermuda, April 21, 1873, 32° 18′ N, 64° 48′ W, to Station 43,
May 1, 1873, 36° 45′ N, 71° 90′ W

HMS *Challenger* left Hamilton on April 21 and soundings confirmed almost immediately that Bermuda was indeed a vast seamount, towering 4 kilometers high from its base on the seafloor. She then shaped a course north and west toward the Canadian coastline under Admiralty orders to investigate the strange anomaly in the thermal structure of the North Atlantic known as the Gulf Stream.

This narrow band of surface water, originating near the Gulf of Mexico and flowing northeast toward the Newfoundland Grand Banks, was known to be much warmer than the surrounding waters. As *Challenger* traversed it, the crew dredged and sounded in the manner to which they had become accustomed, finding that the bottom here was more than 1,500 fathoms (3 kilometers) deep. With their marvelously intricate thermometers and water samplers, they discovered that the width of the stream in that area was fully 60 miles and at least 8°F (4°C) warmer than the waters on either side

The Gulf Stream had been discovered more than 200 years earlier by the Spanish explorer Don Juan Ponce de Leon. Ponce de Leon had been born into the royal court of Aragon and had begun his naval career as a crew member on Christopher Columbus's second expedition to the New World. In 1508 he settled in Puerto Rico and made that island his base for continuing his explorations of the western Atlantic. On March 27, 1513, while searching for a miraculous fountain reputed to confer the gift of eternal youth, he landed on the peninsula that we now call Florida and claimed it for Spain. At the time he did not realize that he had reached the mainland of North America, thinking instead that he had merely found another island. Even so, he named the new land Florida in celebration of its discovery at Easter time (the Spanish name for this festival is Pascua Florida). Returning to Spain, he secured the governorship of Florida and Bimini Island and returned there in 1521. But he was never able to enjoy his newfound dominion, because shortly thereafter he was wounded by a Seminole Indian arrow and died in Cuba. But during his extensive explorations of the Caribbean region he discovered the Gulf Stream and by 1844 the United States Navy had already begun the task of mapping its complex course systematically.

We know today that the name "Gulf Stream" is a misnomer— it implies a simplicity of structure that is not the case. In fact, the Gulf Stream is complex network of surface currents that shift course over time, sometimes disappearing entirely and reappearing elsewhere. The Gulf Stream is an example of a western boundary current, a discrete and narrow surface-water current that is confined to the western boundary of the ocean basins by the interaction of three factors: the general shape of the ocean basin; the rotation of surface waters (clockwise in the northern hemisphere, counterclockwise in the southern) in large-scale circulating features known as gyres; and the action of the wind over large fetches of open sea. The corresponding current in the Pacific Ocean is the Kuroshio, which flows north to about the latitude of central Japan before turning eastward for the open Pacific. In the southern hemisphere (in the Indian Ocean) it is the Agulhas Current. Because of the counterclockwise circulation of Southern Hemisphere gyres the Agulhas Current flows southward rather than northward.

The Gulf Stream is also the surface limb of the current system known to scientists today as the "North Atlantic conveyor" by which warm waters from the Caribbean and the Gulf of Mexico are transported northeastward and then return cool at depth in a current known as "North Atlantic deep water." Nearly all the water that enters the Gulf Stream has been driven westward across the Atlantic by the northeast trade winds.

In the Caribbean and Gulf of Mexico, this stream of surface water is funneled between the continental shelves where its velocity increases to about four miles per hour. This "Florida current" then turns north between the Florida peninsula and the Bahamas and flows north along the line of the North American continental shelf. By the time the current, now the Gulf Stream proper, reaches Cape Hatteras, its velocity has slowed to about one mile per hour, yet its average temperature can be as much as 11°C higher than the surrounding water. The fact that for a thousand miles south of the Gulf Stream the sea surface temperature changes by only 6°C puts this 11°C contrast into perspective.

About 1,500 miles north of Cape Hatteras, the Gulf Stream meets the southward flowing Labrador Current and the confluence of the hot and cold currents causes some of the most widespread fogs in the world. Further out into the Atlantic, the Gulf Stream separates into several different meandering currents that move in the general direction of Europe. In the center of the North Atlantic the Gulf Stream separates into two new currents, the Canary Current that heads south toward Spain and the coast of North Africa, and the North Atlantic Drift that brings warmth and moisture to Great Britain and northern Europe.

The importance of this effect should not be underestimated. In winter the air over Norway is more than 22°C warmer than the average for that latitude at that time of year. Contrast this with the eastern coast of Canada, where the prevailing winds blow out to sea and the Gulf Stream has little effect. Here Halifax, Nova Scotia, a thousand miles further south than Bergen, Norway has an average temperature of –5°C during its coldest month while Bergen remains above freezing.

The lower limb of the North Atlantic conveyor, North Atlantic deep water, is powered not by wind-induced currents but by density-induced sinking. The warm surface waters of the Gulf Stream lose a great deal of moisture through evaporation as they travel north, which has the effect of increasing their density. On top of this, they eventually reach some of the coldest oceans in the world—the Greenland Sea off the eastern coast of Greenland, the Norwegian Sea in an area north of a point midway between Iceland

and Scotland, and the Labrador Sea to the west of Greenland. In winter particularly, these regions are very cold. As the once balmy waters of the Gulf Stream take on the Arctic chill, their density increases further and eventually they sink. This newly formed North Atlantic deep water then starts to flow south again at a depth of more than two-and-a-half kilometers. The North Atlantic conveyor can be thought of as a giant engine for transporting heat from the tropics to the poles, and the Gulf Stream is a vital part of that engine.

It is a paradox that, as we face the risk of global warming, Great Britain and Northern Europe might be cooling. It is already known that subtle shifts in global climate have affected the course of the Gulf Stream, diverting it further south in the direction of Spain. The fear now is that as we warm the atmosphere through emissions of greenhouse gases we might be starting to prevent the northern-latitude waters of the Greenland, Labrador, and Norwegian Seas from becoming cool enough to sink. If that happens, the bottom limb of the North Atlantic conveyor will cease to function properly and the Gulf Stream will once again be diverted south, shifting the climate of Northern Europe at a stroke into the temperature regime more normal for those latitudes.

The Gulf Stream is the thermal blanket that keeps Great Britain warm, and even in 1873 *Challenger*'s crew were well aware of its role in warding off the frigid climate typical of western Canada. As William Spry put it, "Had our shores been without its warming influence, and the British Isles compelled to subsist on their own geographical allowance of heat, we should be left in the same condition." Today we must appreciate that importance again lest the ice arrive and catch us unawares.

Halifax, Nova Scotia, May 9, 1873, 44° 38′ N, 63° 35′ W

Challenger arrived in Halifax, Nova Scotia at noon on May 9, 1873, on a last-minute decision by Captain Nares. He had planned to put in at New York instead, but the Narrative notes, "the usual dirty

weather was experienced on the passage towards New York: occasional strong winds, amounting sometimes to a gale, with light
breezes intervening, and after crossing the Gulf Stream thick fogs,
with rain, until close in to land." Immediately after they finished the
Gulf Stream soundings, the weather deteriorated even further, so
Nares headed instead for Halifax. As *Challenger* steamed gently into
harbor, Joe Matkin found himself noticing how very much like an
English coastal city Halifax was, with its surrounding lighthouses,
forts, and batteries. The weather, too, was more English than they
had become accustomed to in recent months, cold yet bracing; a
welcome change from the stifling heat of the Caribbean and
Bermudas. For Joe Matkin, though, the most impressive sight was at
night, when he watched the northern lights flicker and shimmer in
great shifting veils across the vast Canadian sky.

Halifax was the principal naval station in the British Dominion
of Canada and the dockyard was extensive and well equipped.
Challenger had not sighted the wreck of *Atlantic* on the run in but
the crew soon discovered that it was the principal topic of conver-
sation in the town. Bodies were still being brought in for identification and burial, and the walls of the town's predominantly wooden
buildings were covered with posters, some describing victims still
requiring identification, others offering rewards for information on
those still missing.

Despite the grim business of clearing up in the aftermath of the
tragedy all on board enjoyed Halifax. The food was excellent, "the
fish market more plentifully supplied than any other known, cod,
salmon, halibut and mackerel are very abundant and lobsters only
1d each," wrote Joe Matkin. Taking advantage of this fecundity the
ship's steward, Alfred Taylor, and his assistant, Joe Matkin, took
aboard six months' worth of provisions and 200 tons of coal.

While *Challenger* was in Halifax, a mail steamer arrived from
England, bringing welcome correspondence for all on board and
fresh news of the home media sensation of 1873, the trial of Arthur
Orton, the so-called "Tichborne Claimant." A few years previously

Orton had claimed to be Sir Roger Tichborne, eldest son of the tenth Baronet of Tichborne and heir to the Tichborne Estates. Sir Roger was believed to be have been lost at sea off the coast of Brazil in 1854 but his mother, who had never reconciled herself to her loss, had become convinced that her long-lost son had by some miracle survived. The rest of the family was more skeptical and the case was brought to trial in 1871. It lasted more than three months and in the end Orton was found to be an impostor and sentenced to trial for perjury.

In May of 1873 the start of Orton's second trial was only four months away and the media were in a frenzy of speculation as to whether the original judgment was correct and, in the event of a second conviction, how long he would be sent down for. Feelings on the issue ran high both at home and, as Matkin reported, on *Challenger*, too. Fresh news, especially when fuelled by unregulated shore-based alcohol, was apt to re-ignite fistfights among bluejackets with differing opinions. In time the case of the Tichborne Claimant would become infamous as one of the longest and most expensive trials in British legal history.

Challenger left for Bermuda again on May 19, 1873, having spent only 10 days in Halifax. The expedition was three weeks behind schedule and the Admiralty, which had calculated the duration of *Challenger's* voyage so precisely that all aboard already knew that they were due back at Spithead in April or May of 1876, were anxious that they should make up the lost time before they reached the Cape of Good Hope. Several of the senior officers, including the aristocratic Lord George Campbell, were in New York, having made the 400-mile overland trip by train, and many would not rejoin the ship until she was ready to leave Bermuda for the Azores, several weeks hence.

But there was some compensation for those left behind in the warmth of the welcome they had received in Halifax. As the time for departure came near, the ship was overrun with visitors and their departure from Halifax seemed even livelier than their farewell to Portsmouth five months before. As they steamed slowly out into Halifax Bay to the stirring strains of *Auld Lang Syne* from HMS *Royal Alfred's* brass band, the dockyard walls were crowded with cheering people. *Alfred* had supplied extra men to make up *Challenger's* manpower deficiencies: Five seamen had deserted in Halifax, lured by the promise of instant wealth in the great territories of the United States, while another had been discharged and yet another hospitalized.

The effect of *Challenger's* departure was somewhat spoilt, however, when von Willemoes Suhm, the German naturalist, found that he had mislaid his manservant, and Captain Nares had to send the steam pinnace ashore to find him. The junior officers somewhat irreverently dubbed von Willemoes Suhm "the Baron," and Herbert Swire, the most disrespectful of all *Challenger's* diarists, summed up the general amusement and irritation when he wrote:

> The Baron called for his boots. The Baron is von Suhm and he called for his boots by sending his servant ashore yesterday for a pair just before the ship was about to sail. This was not discovered 'till we had left the wharf under the salute of *Auld Lang Syne* from the *Royal Alfred's* band ... and the sail loosers were actually on the yards waiting for the order to let fall when the Baron remembered that his slave had not returned, so the sails had to be refurled and we secured to a buoy out in the stream, sending the master-at-arms and a sergeant of marines ashore to look for the missing man. This was evidently a disappointment to a number of ladies and gentlemen who had collected in the dockyard and on board the flagship to see us off. Having recovered the servant, boots included, we left in the midst of a smart shower of rain, running out of harbor under all plain sail before a fresh breeze from the northward, and Halifax soon faded in the distance.

After the cold of Halifax *Challenger* was soon once again in the balmy embrace of the Gulf Stream and "the weather was so warm that all the iron in the ship was dripping with damp and the change was considered very unhealthy, a great many are even now on the Sick List with Rheumatics and low fever," Joe Matkin wrote. It would be five months before they saw any cold weather again, as they approached the Great Ice Barrier of the Southern Ocean around Antarctica. To many on board, as they sweltered, that day could not come soon enough.

En route to Bermuda, the *Challenger* company celebrated the Queen's birthday. All hands were issued one-third of a pint of sherry, a drink that was listed wryly in the ship's manifest as "extra surveying stores." For Joe Matkin the day was memorable for another reason: It was his father's birthday. Charles Matkin was still languishing at home in deteriorating health and so the young man drank a quiet toast to his father instead, an offence that, if known, was punishable by death. On May 31, 1873, a pilot took them back through the dangerous reefs that surrounded Hamilton's harbor and they slipped quietly to anchor once again. Wine was issued to all hands again that night. It had been a difficult day and all on board were grateful for it.

No ship in Victoria's navy was as well supplied with wine, pickles, and preserves as *Challenger*. Welcome as they were though, these extra supplies were not given through any spirit of altruism on the part of the Admiralty. No other ship would see so many changes of climate on its voyage, and the Lords in London knew only too well the value of preserving morale. There was, too, the omnipresent threat of scurvy, which could be combated only by a regular intake of vitamin C from citrus fruits. As Matkin wrote, "Captain Cook's expedition round the world about 100 years ago lost nearly half their officers and men from Scurvy." Even in the days of Victoria's sail-to-steam navy no risks would be taken with the sailor's age-old scourge.

TROGLODYTE

The expedition had one final piece of unfinished scientific business to address before it could leave the western Atlantic for the last time. On June 9, 1873, Wyville Thomson and a party of the Scientifics and officers left the ship early in the morning aboard the steam pinnace. Stopping only to pick up the Governor, General

Lefroy, from his mansion on the heights above Hamilton, they headed northwest toward one of Bermuda's most enigmatic natural wonders, the remote, forbidding Walsingham Caves. After an hour's steaming they came to a channel between two narrow necks of land through which the tide rushed with the frenzy of the Severn Bore. When it had ebbed the pinnace slipped quietly through the channel and floated on a vast, glassy sheet of water that glittered like blue glass in the early morning sunlight. They were in the enclosed inland sea known as Harrington Sound, a most peculiar basin, rectangular in shape and about two miles long by a mile wide, that makes up much of the northernmost tip of Bermuda. To the south, a narrow strip of land protects it from the Atlantic while the other three sides rise in richly wooded ridges that form the highest ground in the islands. Completely landlocked, Harrington Sound is a perfect natural harbor and not many years earlier the Lords of the Admiralty had actually considered abandoning their Bermudan base on Ireland Island in favor of it. However, they eventually dismissed the idea on grounds of cost and convenience.

As the pinnace approached the northern shore of the sound, the party could see that the cliffs were pockmarked with a series of low caves that extended all the way to the water line. Such was the stillness of the water that it was impossible to tell where reality ended and the reflection began. In the still, clear air they could see the mats of green algae that covered the cave roofs reflected in the water's quicksilver surface.

The caves are relics from the last series of ice ages when sea level was 125 meters below its current level because so much water was locked up in polar ice sheets. At that time, beginning about one million years ago, the entire Bermuda platform was dry and the island's land mass some 20 times greater that it is today. Although Bermuda today has no natural freshwater reserves of its own, at that time, in the mid-Pleistocene, a substantial body of fresh ground-water made up the bulk of the subsurface. This water eroded the natural limestone of the island into a series of natural caves that became gradually drowned as the ice-sheets melted and the sea level rose.

Today the caves of Bermuda consist of a network of submerged passageways and larger hollows at an average depth of 20 meters below sea level. But it was one of the un-submerged caves, known as Admiral's Cave, that Wyville Thomson had come to see. Fifty-four years before, in 1819, Sir David Milne, at that time commander in chief of the North American and West Indian Station, had discovered a very fine stalagmite growing from the floor of the cave. It measured 11 feet in length, was fully 2 feet in diameter, and weighed three-and-a-half tons. Milne removed it from the cave and eventually it found its way to the Natural History Museum at the University of Edinburgh, the institution where Wyville Thomson was now professor.

In 1863 Sir Alexander Milne, son of Sir David, First Lord of the Admiralty and like his father before him, also commander in chief of the North American and West Indian Station, visited the cave and examined the stump of the stalagmite. He was amazed to see that the stump was regenerating! As he watched he could clearly see drops of water falling from the roof of the cave onto the stump and where they landed two small knobs of calcareous matter were already being formed. He estimated the total bulk of these two small knobs, formed in the 44 years since his father's visit, at about 5 cubic inches. The discovery set him thinking, and he realized that here was, in principle, a perfect way to estimate the age of the original stalagmite itself and from that the age of the Admiral's Cave.

It was his brother, though, David Milne Home, who performed the calculation and presented a paper before the prestigious Royal Society of Edinburgh, estimating the age of the stalagmite and the cave system as some 600,000 years, a figure that sounds plausible given that modern estimates of the cave system's age suggest a maximum of 1,000,000 years. And now, in 1873, 10 years after Sir Alexander Milne's visit, Wyville Thomson, chief scientist of the greatest scientific expedition ever organized, was anxious to examine the stump to see for himself how much more it had grown. But when the Scientifics examined the stump, they found that the tiny calcareous lumps had not perceptibly grown since Sir Alexander Milne's visit.

Despite this disappointment, all agreed that the caves of Bermuda were magical. In all the caves they were surrounded by beautifully fluted and fretted columns whose pure white frosted surfaces shone out like beacons in the harsh magnesium light of their lanterns. But it was Painter's Vale Cave that induced their greatest awe. At the foot of a bank of debris lay a pool of deep clear water, perfectly still and reflecting the roof like a mirror. Clambering down the slope, as their eyes became accustomed to the dark, they could see that the lake stretched far back into the gloom. A little punt was moored near the shore and, lighting candles, Nares rowed Governor Lefroy back into the darkness. From the narrow shore the party watched as the dim light of their candles receded and dimmed, their voices became hollow and distant, and they became denizens of the underworld—troglodytes.

They were not alone in enjoying this subterranean existence because Bermuda's caves support a diverse fauna specially adapted to a lightless existence. Today we know that the fauna of Bermuda's caves represents what is known as a biodiversity hotspot. Biodiversity hotspots are defined as small areas with exceptional concentrations of unique species. Bermuda's caves qualify as such because of a very rich community of cave-dwelling animals known formally as anchialine organisms. Two orders, one family, and fifteen genera of crustacea—more than 60 species in all—from the caves were recently identified and they are all new to science.

These animals, often without eyes or pigmentation, neither of which is needed in the lightless world of the caves, are highly endemic; that is, they are restricted to a narrow strip of land between Harrington Sound and Castle Harbor; the north shore that Wyville Thomson and the Scientifics visited that long ago day in 1873. Many of the cave-dwelling animals are even more restricted, some being unique to individual caves. They get their nutrition from planktonic organisms brought in on the sluggish tides where the caves connect with the ocean. These tides slowly replenish the deep fully marine waters with oxygen and nutrients while the surface waters, containing as they do a component from rainfall, remain brackish. Strangely enough, although it was not immediately apparent to the Scientifics, the caves of Bermuda hold one of the very things that had motivated the *Challenger* expedition in the first place—living fossils.

Some of the cave's animals have organs and structures that are known only from the fossil record. On top of this, many of Bermuda's cave-dwelling species are very similar to those known from Europe and Asia and must have arrived on the island before the spreading of the mid-Atlantic Ridge separated Bermuda from those continents.

Other animals in Bermuda's caves seem to be close relatives of cave dwellers known from isolated islands elsewhere in the Atlantic and, surprisingly, in the Pacific, too. Still others seem to be related to species known only from the deep sea. The question of how animals from such diverse habitats arrived in the caves of Bermuda is currently the subject of intense scientific research that is helping to answer broader questions about the evolution of oceanic species.

Despite the great antiquity of Bermuda's cave system and its inhabitants, its ecology is extremely delicate and currently under severe threat from that most rapacious of all predators—man. This is worrying because of the uniqueness of the ecology. The huge impact of man on the caves, particularly in the last century, has been due mainly to construction activities and water pollution. The U.S. military has poured raw sewage into the pool of Bassett's Cave, which as long ago as 1837 was known to be the most geologically interesting of all the caves. Other threats come from dumping, littering, and, tragically, from deliberate vandalism. Quarrying operations have already destroyed many caves and have ramifications further afield when the water of adjoining caves becomes polluted.

In such cases the death of the fauna is inevitable and because many animals are cave-specific this means extinction—the final and absolute removal of a species from our planetary ecology.

For Herbert Swire, the caves were only moderately interesting. He found it hard to share the Scientifics' enthusiasm and treated the cave expedition as an excuse for a picnic. But for Swire the perfect picnic included a crucial ingredient, one that was sadly lacking that day at the Walsingham Caves: there were no ladies. Swire was a young man and felt the lack of female companionship especially keenly. He was also incurably romantic and not long after wrote in his diary, "I have just finished reading the late Lord Lytton's last work, published since his death, *Kenelm Chillingly*. What a sad ending to a noble book! I think I have never been so near blubbering as I was when I came to the chapter in which Kenelm receives from Mrs. Cameron, Lily's last message, in the shape of a note written before she died. I wonder has such a truly noble girl ever lived in reality upon this earth?" For Swire, though, as for the others, it would be many months before he had the female companionship he so desired, because the Azores, the Cape Verde Islands, and the vast South Atlantic beckoned.

Their final act before leaving Bermuda was to erect the gravestone they had brought back from Halifax in the naval cemetery there. It was a large marble cross and bore the inscription "This stone is erected by the Officers and Crew of HMS *Challenger*, to the memory of Adam Ebbles, Naval Schoolmaster who died at Bermuda, April 4th: also to Wm H. Stokes, 1st-Class Boy, who was killed March 25th 73, off the West India Islands. In the midst of life we are in death."

Kelp and Cold Light

Hamilton, Bermuda, 32° 18′ N, 64° 48′ W Atlantic Ocean, to Edinburgh, Tristan da Cunha, 37° 03′ S, 12° 18′ W

Nearly everyone was pleased to leave Bermuda. Herbert Swire wrote "Thank heaven we sail tomorrow! This is indeed a most fearful place to be stationed. Oh Bermuda, I regret thee not at all. Thy shady lanes winding between groves of magnificent but mosquito infested oleanders; thy cedars that give no shade, and thy burning sun from which there is no escape. . . . May I never clap eyes on thee again, and mayst thou soon declare thy independence and free the English nation from the obligation of occupying thy fortresses."

Joseph Matkin, too, was not sorry to see the back of Bermuda but he was not able to properly enjoy his departure. By the time *Challenger* cleared her moorings at 3 P.M. on June 12, 1873, he was doubled over with racking stomach pains. He was violently sick, unable to keep anything down and suffering chronic diarrhea. As if to punish him for the dislike he felt for the islands, their parting gift to him was the travelers' disease, dysentery. He dosed it with arrowroot and Liebhes beef tea, a concoction, invented by a German chemist of the same name, which had rocketed to fame in the Victorian world as a universal panacea. Indeed some were so convinced of its efficacy, they predicted that it would eventually replace the need for food. The bluejackets, though, had their own name for it—"animal fluid"—reflecting their cynicism at the newfound love affair that the middle classes were enjoying with the new Victorian religion, science.

Challenger left Bermuda in a hurry, driven from there by news that the disease was spreading across the island. But, as Matkin could now unhappily testify, they were not completely successful in avoiding it. The crew complement was further reduced by the need to leave two particularly severe cases behind in hospital. This loss, combined with the desertions in Halifax, had brought the crew complement down to 236. All the missing men would have to be replaced at the Cape. But first there was the long haul down the Atlantic via the Azores, Cape Verde Islands, St. Paul's Rocks, Bahia, and Tristan da Cunha, a journey that would take almost four months. Yet on the first leg to the Azores dredging proceeded apace and successfully, the Scientifics were gleeful, and Captain Nares regularly issued wine to *Challenger's* hardworking crew.

It was in Bermuda that Wyville Thomson started the first of the several articles he was to write for the popular magazine *Good Words*, despite a correspondence workload that was already high. He was already sending regular scientific updates back to the *Philosophical Transactions of the Royal Society of London* as well as to the Macmillan brothers recently established journal, *Nature*. But Wyville Thomson, like many other great Victorian thinkers (including his estranged friend William Carpenter and the titanic Thomas Henry Huxley, future president of the Royal Society), felt the need to promulgate the excitement of science to a more general audience. The *Challenger* expedition, being just about the biggest thing ever to happen in Victorian science, had attracted a lot of general interest, and before he left London, Wyville Thomson had agreed to contribute to *Good Words* because of its wide readership.

While in Bermuda, and despite the fact that they had already been three months at sea, he started his account with the tale of the shakedown crossing from London to Madeira. He then wrote lyrically about the caves and corals that they had just seen in the West Indies. As the ship pulled away from Bermuda, northeast toward the

Azores, he explained to his audience the importance of their new destination; situated on the northern border of one of the strangest places in the North Atlantic. For more than 300 years, since the time of Columbus, its name had struck terror in the hearts of mariners all over the world and tales were told of ships becalmed for all eternity in a choking mass of seaweed that matted the center of the mid-Atlantic. When *Challenger* finally made harbor in the Azores, she would have sailed completely around the Sargasso Sea.

THE PRAIRIE OF KELP

Strangely enough, legends of "a sea of lost ships" were common centuries before the Bermuda Triangle became notorious. Since the Middle Ages, floating derelicts have often been found in this region of the Atlantic, which broadly extends between about 20° N and 35° N and 30° W and 70° W (the large uncertainty in this estimate is part of the mystery of the Sargasso Sea). The legend maintains that the Sargasso Sea derelicts are found shipshape but otherwise bereft of a living soul. On one occasion a slaver was sighted but when boarded was found to contain nothing but the skeletal remains of crew and slaves. In 1840, the merchant ship *Rosalie* sailed through the area but, as the London *Times* later reported, was thereafter found drifting and derelict. In 1857, only a handful of years before the *Challenger* expedition, the bark *James B. Chester* was found becalmed in the Sargasso, with the chairs upended, a putrefying meal still on the mess table, and no sign of the crew.

Even after the *Challenger* voyage, legends about ship disappearances continued to haunt the area. In 1881, the schooner *Ellen Austin*, bound for London, discovered a derelict adrift in the Sargasso. The captain put a prize crew aboard but then the two ships became separated by a squall. When *Ellen Austin* re-sighted the derelict, the prize crew was gone. And today, in the early twenty-first century, more recent legends of the Sargasso continue to haunt us. As recently as 1955, the *Connemara IV* was found deserted and drifting in the area, only 150 miles from Bermuda.

For hundreds of years the Sargasso Sea, like the Bermuda Triangle, has been a magnet for the tabloid press. Lurid nineteenth-century paintings show sailing vessels being devoured by the weed that floats on the surface of the sea: *Sargassum*, so named by Portuguese sailors who spotted the resemblance of the weed's air-filled bladders to the grapes of their homeland. And like the myth of the Bermuda Triangle, the legends of the Sargasso Sea have some basis in fact. Much of the sea's peril comes from its location in the almost windless "Horse Latitudes," so called because ships en route to the Spanish Main were often becalmed there and their horses were slaughtered for fresh meat and to preserve water.

The Sargasso Sea is surrounded by some of the strongest surface-water currents in the world: the Florida Current to the southwest, the Gulf Stream to the northwest and north, the North Atlantic Current to the north and northeast, the Canary Current to the east, and the North Equatorial Drift running along the entire southern margin of the sea. These currents form a cordon around the sea, isolating it from the rest of the Atlantic. This isolation causes two other curious features: the sea's unique temperature structure and its unique ecology. The Sargasso Sea is actually a thin lens of warm water perched on top of much colder water and is home to great floating beds of the *Sargassum* kelp that gives the sea its name. Christopher Columbus noticed this unusual plant on his first trans-Atlantic voyage in 1492. He encountered great floating masses of it not far to the west of the Azores and by the time he reached the middle of the Atlantic there was "such an abundance of weed that the ocean seemed to be covered with it." When his ship was becalmed for three days, his men grew alarmed, thinking that they had reached a shore and were in imminent danger of running aground. To allay their fears, Columbus sounded and found the bottom to be well below the maximum depth his line could reach. We know now that the ocean floor beneath the Sargasso Sea is at a staggering depth of more than three miles. The silent landscape there is named the Nares Abyssal Plain in honor of *Challenger's* captain.

In late Victorian times the legend of the Sargasso Sea, like many macabre tales, continued to grow. As late as 1897, and in the aftermath of the *Challenger* expedition, a writer in the *Chambers Journal of Popular Literature, Science and Arts* said, "it seems doubtful whether a sailing vessel would be able to cut her way into the thick network of weeds even with a strong wind behind her. With regard to a steamer, no prudent skipper is ever likely to make the attempt for it certainly would not be long before the tangling weeds would choke his screw and render it useless."

Because of variations in the strength and direction of the sea's boundary currents, the borders of this prairie of kelp are not constant. Indeed attempts to define the Sargasso Sea's area have always proved difficult. Only five years after *Challenger* returned home, the German scientist Otto Krummel analyzed the reports of German sea captains, who for many years had been required to record their observations of the location of the drifting weed beds of the Atlantic. Krummel concluded that the Sargasso Sea covered an elliptical area of some 1,720 square miles, extending from the mid-Atlantic to the North American coast.

In 1923 the Danish botanist Otto Wing made a second attempt to define the boundaries of the Sargasso and his estimate dwarfed Krummel's. He placed the eastern boundary near the Azores, almost exactly where Columbus first encountered floating weed 500 years before, and the southern boundary near the West Indies. The western and northern boundaries, Wing found, shifted position on an apparently seasonal basis, which he attributed to yearly changes in local weather patterns. But not until the 1930s and the 1940s did oceanographers realize that the Sargasso Sea was better defined in terms of its encircling currents than by any other criterion. This explains how the precise size and shape of the sea varies according to the position of these boundary currents.

Later investigations focused on the sea's thermal structure and determined that it is actually a huge body of warm water separated from the colder layers below by a pronounced thermocline—a zone of rapidly changing temperature. In profile, the sea is lens-shaped because of the current-driven clockwise rotation of the Sargasso water mass. This clockwise rotation piles water up in the center of the sea until it is fully two feet higher there than at the outermost edges. At its deepest, the Sargasso lens is only about 3,000 feet thick, so it can be thought of as a thin homogenous layer of warm water lying atop a body of cold water at least five times thicker. Temperature measurements also defined the edges of the sea more accurately and confirmed the importance of the encircling surface water currents.

But what keeps the vast mats of kelp—which is, after all, the best known feature of the region and the source of its legends—trapped there? In 1927, while voyaging across the Atlantic, physicist Irving Langmuire noticed that when the wind veered at right angles to its former direction, the bands of seaweed reformed in the new direction within 20 minutes. From this observation he concluded that something more than simple wind power must be responsible for the orientation of the bands, especially because the winds of the Sargasso Sea are, as Columbus discovered, notably weak. Langmuire suggested that shifts in small surface-water currents within the sea might be responsible for the orientation of the bands. He later demonstrated experimentally that the action of even sluggish winds over open fetches of water produces long avenues of counter-rotating eddies with bands of sinking water between them. This sinking region concentrates the *Sargassum* into long characteristic bands.

So what exactly are these weeds that have excited so much interest, controversy, and fear over the centuries? Between 1932 and 1935 Albert E. Power, director of the Bingham Oceanographic Laboratory at Yale University collected more than 5,000 pounds of the *Sargassum* kelp and made a detailed study of its biology. He found that more than 90 percent of the *Sargassum* was made up of only two species, both unique to the Sargasso Sea. They are never found attached to anything and they lack the organs for sexual reproduction.

So where do the weeds come from and how do they reproduce? Columbus suggested that the drifting clumps were torn loose from great submerged beds near the Azores and then gradually collected in the Sargasso Sea. However, no such beds have ever been found either in the Azores or even in Bermuda, which lies more or less in the geographic center of the sea. More recently, botanists have suggested that the weed originates in the great floating banks of kelp found near the West Indies and in the Gulf of Mexico. Yet there is no evidence that these banks can spawn the presumably huge streamers of weed that would be expected to be found traveling northwest via the Gulf Stream. Also, detailed ecological studies have shown that the sea's standing crop of weed weighs a staggering 7 million tons, too much to have been brought north on the Gulf Stream. Finally, the *Sargassum* does not look moribund and decrepit, as it would if it really had been torn loose from a colony and transported thousands of miles. On the contrary, the weed is a healthy green color and routinely shows evidence of fresh new growth.

In short, the evidence suggests Sargasso Sea kelp is a native of the sea itself. Although its ancestors might have been bottom-dwelling kelp, over the course of eons it independently evolved the ability to prosper and reproduce in its surface environment. Today *Sargassum* reproduces asexually by vegetative budding of new shoots that eventually break off to form new plants.

This unique floating world of kelp also supports an ecosystem of animals—crabs, mollusks, and fish, many of which have evolved bizarre camouflage that makes them indistinguishable from the kelp on which they live. Perhaps the best example is the pipefish *Syngnathus pelagicus*, a close relative of the common seahorse that has evolved flaps of greenish-brown skin that perfectly disguise it as a frond of *Sargassum* kelp. In all, there are more than 50 species of fish, and many more species of invertebrates (such as gastropod snails, moss-like bryozoans, polychaete worms, anemones, and sea-spiders) whose lives are intimately linked to the ecology of this prairie of kelp.

Another unique feature of the Sargasso fauna is that they are omnivores, capable of deriving nutrition from vegetation as well as animal prey. This generalist ecology is a reflection of the strangest feature of all in the Sargasso Sea: In biological terms it is a desert.

Like the rest of the world's seas, the Sargasso Sea has a population of the subsurface algae called coccoliths as well as planktonic foraminifera. But strangely, despite its vast floating forests of kelp, the Sargasso Sea is the least productive of any sea in the world. This is clearly seen in the extreme clarity of the dark-blue waters between the kelp bands; in other oceans these waters would be tinged green by the teeming photosynthesis that is a feature of most brightly lit upper-ocean waters. This is a paradox; how can we explain the fact that despite its obvious and abundant surface life, at even shallow depths the Sargasso Sea is, comparatively speaking, biologically barren?

The most likely explanation relates to the most obvious and earliest known facts about the Sargasso Sea: Things that enter it do not leave. This is a function of the bounding surface-water currents and the slowly rotating mass of warm water in the middle, which has the effect of inhibiting the supply of nutrient so that photosynthesis cannot proceed with its customary vigor. This physical isolation is further compounded by the sea's thermal isolation, which prevents nutrients from welling up from below.

The Sargasso Sea is one of nature's oddities and sadly, it has its own unique pollution problems. The same circulation phenomena that concentrate the *Sargassum* kelp in the center of the sea also attract garbage from all over the North Atlantic, from trash thrown overboard by ocean-going liners to the tar balls that result from oil spills. The tar balls in particular are a problem, because they are not biodegradable and they do not sink. Gradually, the Sargasso Sea is becoming the world's biggest oil slick and one of the concerns for our, and future, generations must be how to save this unique ecosystem.

The Azores, June 30, 1873, 38° 30′ N, 28° 00′ W

On July 1, 1873, *Challenger* dropped anchor off Horta, capital of the Azorean island of Pico. "We are pretty close in now and in a few minutes the pipe will go 'all hands bring ship to anchor'" wrote Joe Matkin, "and out will come the Portuguese boats with fish, fruit etc.... the chief town [is] Horta, a very pretty place built all along the beach and looking like Brighton from the sea." For the young scribe below decks it was all a welcome change from the heat, humidity, and disease that had dogged their stay in Bermuda. After a considerable delay, the quarantine boat came alongside. Eventually they were given grudging permission to land but they soon found that the town was in the grip of a smallpox epidemic and that another raged in Madeira. On the strength of this, Nares decided that their visits to both islands would be as short as possible and that they would make all plain sail for the South Atlantic and the Cape at the earliest opportunity.

Moving swiftly on from the contaminated island, *Challenger* put in at St. Miguel on July 4, 1873, American Independence Day, for some much-needed shore leave. Despite the 18-day transit time, the trip from Bermuda had been an enormous success and the Scientifics at least were in fine fettle. To the middle and upper classes of Victorian Britain the Azorean island of St. Michael was a favorite stopover, being the largest of the islands and famous for the sweetness and quantity of its oranges. The bluejackets, though, favored the Azores for a different reason: They were only 1,100 miles, five day's sailing, from Southampton. To the tars of Victoria's navy, especially those returning from the farthest flung corners of the empire, the Azores were the gateway to home. And here was the rub, the reason for the bluejackets' restlessness. Until a few days ago it had seemed that *Challenger* would have to make that 5-day journey home for emergency repairs. The condenser that converted sea-water to fresh water—providing drinking water for the crew—had malfunctioned and fixing or replacing it at the Portsmouth dock-yard had seemed like a real prospect.

In fact, the condenser had not operated properly since their arrival in Bermuda, but in the true tradition of the greatest navy in the world, Spry and the other engineers had managed to fix it at sea, with the result that now the much anticipated trip home would not materialize. *Challenger* would, after all, be turning immediately southward on the long transit to Simonstown and the Cape of Good Hope. Matkin realized sadly that whatever happened now it would be another three years before they saw the Azores again. Only then, with a fair wind up-channel, would they truly be only "five days from home."

There was also the lingering suspicion below decks that Captain Nares's reluctance to sail for home had as much to do with the many desertions that had already occurred as the skill of William Spry and his fellow engineers. Perhaps Nares thought that if they did touch home soil again, he would lose most of his crew. Certainly their brief stopover in Halifax had hemorrhaged their manpower. With these bitter thoughts swirling through their minds, it was inevitable that as soon as they hit shore many of the sailors found themselves in brawls with the locals. On July 5th, some of the tars came back on board with the makings of a Sunday dinner that they had "liberated" from the islands—half a sheep here, several pounds of goat meat there, some even carrying pigs' heads. To a man they were drunk and had been causing trouble ashore. Joe Matkin, with his moderate tastes, watched in amusement as his messmates came staggering and slipping up the gangplank and tried to form coherent replies to the enraged officer of the watch. Those that could not walk in a straight line were written up as being drunk, fined a day's leave, and sent below to sleep it off.

In the end Matkin was glad not to go ashore until the 7th. The

fallout from the brawls caused the Portuguese authorities to com-
plain officially to *Challenger's* acting officer (Nares, the Scientifics,
and several of the officers including Herbert Swire, having headed
off into the interior of San Miguel to investigate the hot springs
there) and for a little while the matter was rather tense. The next
day, a Sunday, there was a repeat of the trouble of the previous night.
The Azores being a Catholic country, the wine shops and bars
opened after Church at midday and the tars drank the strong red
wine as though it were the weaker ale of England. This time the
regiment was called out, backed up by several enraged civilians, and
the fighting was intense. The soldiers used their bayonets and several
men returned to the ship with cuts and gashes. But they dressed
each other's wounds so that the ship's surgeon would not have to
become involved and "write them up" again. However, because
several Portuguese civilians and soldiers were injured, the bluejackets
could not disguise the trouble completely and after this second
incident, shore leave was cancelled for all except the very few who
had not already been ashore.

    Challenger left San Miguel on July 9, 1873 under all plain sail
for the island of Madeira, 490 miles to the southeast. The transit
would take a week because of the usual delays of sounding and
dredging. Because of the smallpox outbreak there, none aboard
expected to stay very long in Madeira and indeed, Nares was so
impatient to leave that the ship, which arrived in the evening of
July 16, was in port only 24 hours before departing again for the
Cape Verde Islands, a thousand miles to the south.

    Challenger lingered in Madeira just long enough to take on
board wine, fresh meat, vegetables, bread, and fruit. The brevity of
the visit was a bother to the crew because it meant that the mail bag
was barely opened before it was shut again and those who were
tardy in composing their letters home had missed their chance until
Challenger arrived in the Cape Verde Islands at the beginning of
August. But Joe Matkin was not too unhappy; that brief stopover

had brought him eight letters from home as well as the *Mercury*, the local paper from the tiny English county of Rutland where he grew up.

Under the influence of the northeast trade winds, *Challenger* made good time despite the frequent stops required by the Scientifics. Heading south, they dredged again near the Canaries in an effort to rediscover the area where they had found the manganese nodules on their first trans-Atlantic transect. They found the narrow plateau where they had retrieved the first specimens but not the exact spot, in spite of having, as George Campbell put it, "the night before run before the wind under bare poles so as not to overshoot it."

Following the 1,000-mile run south to the Cape Verde islands, by July 27th the *Challenger* crew was dredging off the island of St. Vincent. There they were to pick up two new members of the crew, a new sub-lieutenant and a replacement schoolmaster sent out from England to take over the care of Captain Nares's young son, Billy, following the death of Adam Ebbels in Bermuda. But when they arrived at St. Vincent they found that the schoolmaster had disappeared. The new lieutenant, Harston, could shed little light on the matter except that the schoolmaster had gone out for a walk soon after their arrival eight days before and had not been seen since. Enquiries of the British consul on the island elicited little except the disquieting information that the new man had arrived penniless—his pay was ready for him aboard *Challenger*—and the fact that the consul had declined to advance him any money while he waited for the arrival of his new berth. A search revealed nothing and, with the tight timeline imposed on the ship's itinerary by the Lords back in London, *Challenger* could not tarry.

Despite heavy hearts at not being able to locate the school-master, the departure from the island was a welcome relief. "St. Vincent is, I think, the most dismal place I know," growled Campbell, while Swire wrote, "I was not much grieved to get away from St. Vincent. It was indeed but a sorry hole." The word among the crew was that the other islands of the group were much more pleasant, but Joe Matkin was not about to be easily impressed, noting his opinion of the island group even as they left St. Vincent "they are rocky, barren and unfruitful, scarcely worth occupying." During their survey of the harbor at St. Vincent, the troopship *Simoom* arrived from Britain bound for South Africa and the fresh trouble in the colony there, the Ashanti War. The King of the Ashanti tribe had declared war against the British Empire and had marched on a settlement with 30,000 men. They had been repulsed at the Battle of Elmina by a lieutenant and 30 bluejackets, but now the *Simoom's* contingent of marines were outbound to administer some rough colonial justice, or as Swire put it bluntly, "to slaughter Ashantees at Cape Coast Castle."

Matkin's misgivings about the rest of the Cape Verde islands were assuaged when on August 6th *Challenger* anchored at the island of St. Iago (Santiago). All on board found it to be such a pleasant change that they elected to stay for three days instead of the one that was originally planned. Fruit was abundant so they stocked up on the sailors' traditional defense against scurvy, taking on board no fewer than 5,000 limes and a considerable quantity of oranges, bananas, and coconuts. The coconuts in particular impressed Swire: "The milk from the green coconuts is very refreshing on a hot stifling day, and differs very much indeed from the liquid got from the nuts in the state in which they arrive in England."

From Santiago, *Challenger* headed south and east into the Gulf of Guinea, that expanse of ocean under the bulge of Africa. Their course took them far eastward of the usual track of ships bound for the southern hemisphere, a direction that was not welcomed by the crew. "The Gulf of Guinea is no pleasant place to be in," wrote Swire, "because of its horrible, wet muggy atmosphere and the frequent rains which occur all the year round." Their proximity to Sierra Leone and the "White Man's Grave" was a cause of considerable consternation for many of the Scientifics because of the life insurance policies that would be invalidated if they went to the west coast of Africa. But Nares knew what he was about, and having approached within 160 miles of the coast, "till the third degree of north latitude," as Campbell put it, turned about and headed westward for the desolate peaks in the equatorial Atlantic known as St. Paul's Rocks. On this leg of the journey Campbell saw something that affected him more deeply than any other sight he had seen in nine months at sea.

> On the night of the 14th the sea was most gloriously phosphorescent, to a degree unequalled in our experience. A fresh breeze was blowing and every wave and wavelet as far as one could see from the ship on all sides to the distant horizon flashed brightly as they broke, while above the horizon hung a faint but visible white light. Astern of the ship, deep down where the keel cut the water, glowed a broad band of blue, emerald-green light, from which came streaming up or floated on the surface, myriads of yellow sparks which glittered and sparkled against the brilliant cloud-light below, until both mingled and died out astern far away in our wake. Ahead of the ship, where the old bluff bows of the *Challenger* went ploughing and churning through the sea, there was light enough to read the smallest print with ease. It was as if the milky way . . . had dropped down on the ocean, and we were sailing through it.

Challenger had discovered yet another carpet of the deep, this one powered by a phenomenon that is common to all the world's oceans, bioluminescence.

THE LIGHT OF THE OCEAN

It is important to understand that there is an essential difference between the phosphorescence mistakenly mentioned by Campbell and true bioluminescence. Phosphorescence is a physical process like fluorescence, in which an electron orbiting the nucleus of an atom is driven temporarily into a higher orbit when hit by a passing photon (particle of light). This energy is almost immediately re-emitted (at a longer wavelength; that is, one with less energy) in the form of another photon. Phosphorescence is similar, except that the time taken to emit the energy is greater, which is why light shines from phosphorescent materials, such as glow-in-the-dark stickers, even after the initial source of light energy is extinguished.

Bioluminescence, in contrast, is a true biological photochemical reaction. Contrary to popular belief, bioluminescence is very common in the sea, and indeed is the main source of light in many parts of it. It is found in many different groups of animals and plants and appears to have evolved independently several times. All bioluminescent systems are based on two chemical compounds: One, which actually produces the light, is called the "luciferin"; the other, a "luciferase," is the biological catalyst that controls the reaction. (It is a convention in biology that compounds ending in "ase" are some form of biological catalyst known formally as an enzyme.) Although the precise chemical structure of luciferins and luciferases varies between groups of bioluminescent organisms, four combinations are common: those in bacteria; those in the seed-shrimp *Vargula*; the "Coelenterazine" system (the most widely used in the animal kingdom); and that used by the fantastically abundant, floating, single-celled plants known as dinoflagellates. The dinoflagellates's bioluminescence is generated by a modified form of chlorophyll, which caused the "phosphorescence" that so impressed Campbell that long ago night in August 1873.

All these systems operate the same way even if the chemical basis varies. The luciferase adds a molecule of oxygen to the luciferin, which stimulates it to emit a photon of light and creates a new compound called an oxyluciferin. Because of this chemical transformation, the luciferin must be regularly replaced either through synthesis from more basic chemical building blocks or through the emitting organism's diet. Sometimes the luciferin and the luciferase are bound together, forming a larger molecule called a photoprotein, which emits light when a negatively charged atom or ion, often calcium, is added. In all cases, however, bioluminescence is unique in that it is a "cold light." Unlike most artificial light sources, which depend on the glow of heated materials, or even natural light sources like the sun and stars, bioluminescent light generates very little heat radiation.

The light that Lord George Campbell and the others saw en-route to St. Paul's Rocks was predominantly green. This is because bioluminescent light is concentrated toward the blue end of the visible electromagnetic spectrum—the wavelength of light to which seawater is most transparent. The dinoflagellates that provided the illumination at the bow and the keel as *Challenger* ploughed through the water produce bursts of illumination that last only a tenth of a second, but each burst consists of 600,000 photons. Larger organisms such as jellyfish produce much larger numbers of photons, often billions, in bursts that can last tens of seconds. In single-celled organisms like the dinoflagellates, the luminescence is triggered when the cell surface is deformed by minute forces such as the disturbed currents of water from a passing ship. The deformation of the cell surface translates into a tiny electrical current that admits subatomic particles called protons into the interior of the cell and stimulates its luciferin-luciferase system. In higher multicellular organisms though, the luminescence is under nervous control and is often triggered by some behaviorally significant event. In other organisms light can be produced by "photic excitation," that is, the receipt of a photon of light from an outside source, even another luminescing organism. This can lead to a cascade of luminescence among shoals of organisms after just one has been mechanically or behaviorally stimulated.

But what are the advantages of bioluminescence? They clearly must be major because it is estimated that bioluminescence has evolved more than 30 separate times in different, and unrelated, groups of animals. It is thought that there are two main benefits: defense and mating. The defense idea is based on the "burglar alarm" hypothesis. The idea is that if a prey is endangered by a predator, it will bioluminesce, thereby drawing attention to itself. If this attracts a larger predator the chances are that it will be interested in something larger than the original prey and attack the original predator, thereby giving the bioluminescing prey the opportunity to escape. The mating idea is more obvious. It is thought that in many higher invertebrates the ability to bioluminesce enables mates to find each other in the darkness of the ocean's depths. In some cases the light emitted for mating might be of a species-specific frequency, enabling mates to find each other without alerting potential predators.

St Paul's Rocks, Equatorial Atlantic Ocean, August 27, 1873, 01° 00′ N, 29° 23′ W to Edinburgh, Tristan da Cunha, South Atlantic Ocean, October 15, 1873, 37 ° 03′ S, 12° 18′ W

"On 27th August," wrote Lord Campbell,

> ...we sighted St. Paul's Rocks, steamed to leeward of them, and as there is no anchorage, sent boats with ropes and hawsers to the rocks, wound a rope round and round a bit of rock, made a hawser fast to the rope and swung to it with a length of 75 fathoms of hawser, 104 fathoms of water under our bows and there we comfortably lay for a day and two nights, made fast to a pinnacle of rock in the middle of the Atlantic!—something no other ship has ever done here before. St. Paul's Rocks are a cluster of five separate craggy rocks, all lying close together in a horseshoe shape, the highest being about 60 feet high, which, as are also two other peaks rather less high, is colored white from the birds 'boobies' and 'noddies' which were sitting about on the rocks, flying over the ship and close over the sea, in thousands.

Bringing the big ship to station at St. Paul's Rocks (see Figure 6), with its sheer walls of rock and no mooring, was tense work, and Captain Nares and Navigating Lieutenant Tizard took no chances as they guided the corvette close in to the island. Together they stood in the foretops and conned the ship in through the seething maelstrom of the equatorial current. "I never properly realized the strength of an oceanic current," wrote Moseley, "until I saw the equatorial current running past St. Paul's Rocks . . . a small fixed point in the midst of a great ocean current, which is to be seen running past the rocks like a mill race. . . ." Indeed, so strong was the current that *Challenger* was held secure and immobile with just a single hawser. The remoteness of their situation was not lost below decks, either. Joe Matkin wrote of the Rocks, "They are 850 miles from the African, and 650 miles from the American continent, and are only 90 miles from the equator. The sea all around them is two

FIGURE 6 *Challenger* at St. Paul's Rocks

miles in depth, they rise only 60 feet out of the water, and as it breaks all over them, landing is very difficult. . . . They are out of the track of any ships, and as nothing is to be obtained, no vessel ever comes, the last known to call here was a man of war in 1845."

By 6:00 that night they had made fast and a jolly boat was sent out to attempt a landing on the island's sheer sides. This, too, was no easy matter, with the surf rising and falling a full 7 feet up and down the glistening black sides of the island. Eventually, landing was accomplished by "a spring and a scramble when the boat is on the top of a wave," as Wyville Thomson put it, and "When we landed the sun was just setting behind the ship. There was not a cloud in the sky, and the sun went down into the sea a perfect disk, throwing wonderful tints of rose color upon the fantastic rocks."

In fact, St. Paul's Rocks had excited the interest of no less a person than the *Challenger* expedition's spiritual and intellectual mentor 31 years earlier. Charles Darwin was a young man of 23 when *Beagle* paused there on its long voyage south. Darwin was curious that the rocks of the islands were not of volcanic origin, as was the case for almost every other oceanic island, and he surmised that St. Paul's Rocks were the tips of mountains of a much larger land mass, perhaps even a continent, that had sunk beneath the Atlantic waves eons before. *Challenger's* Scientifics agreed and hypothesized that the next leg of their journey would yield evidence that they were indeed in shallow water, and following the course of an ancient sunken continent.

For Wyville Thomson, it was another moment of delicious anticipation in a voyage that so far had been just one scientific triumph after another. Only a couple of days earlier he had to deliver on a foolish bet that he made after one of the earliest successes of the voyage. On the transit from the Gulf of Guinea to St. Paul's Rock's, they had dredged another specimen of *Umbellularia*, a crinoid of the same genus that had excited so much interest before their arrival in Gibraltar because it seemed to be the first proof of Darwin's suggestion that the deep ocean was a haven for "fossil" life. To most of the officers aboard *Challenger*, though, that second discovery had a more immediate and welcome consequence than merely adding to the annals of science. "The Professor having previously made a bet with the mess generally that we should not get another *Umbellularia* tonight at dinner paid the penalty in champagne 'all round'" wrote Campbell.

On August 29, 1873, *Challenger* cast off from the desolate pinnacles of St. Paul's Rocks and headed for the equator, only 90 miles distant. Matkin wrote, "We crossed the Line at 1pm on Aug. 30th, no shaving etc was allowed," (a reference to the traditional treatment by old hands of those who crossed the equator for the first time), "but the Captain issued Wine to all hands in honor of the event."

Their next stop, the island of Fernando Noronha, hove into view on September 1st. Brazil used the island as a penal colony for its worst offenders, "The prisoners in this establishment are chiefly of low grade," wrote Wyville Thomson, "and most are convicted of

heavy crimes. . . ." The landing at Fernando Noronha was difficult
because of the heavy swell, and the boat containing Henry Moseley
and Herbert Swire capsized. All on board were dumped unceremo-
niously in the water and two were trapped underneath. As Swire
and Navigating Lieutenant Tizard fought among the breakers to
free the trapped men, Swire could not contain his disgust at the
actions of the enlisted men. He wrote later, "All the bluejackets
made for the shore and left Tizard and myself to extricate two
people who were still under the boat, but on finding that there was
no danger (the water was only about five feet deep) they [then]
came gallantly to the rescue."

Meanwhile, Wyville Thomson and Nares (accompanied by his
young son Billy) were paying a visit to the governor of the island, a
major in the Brazilian army. They were courteously received. "We
found the Governor a grave, rather saturnine Brazilian, silent, partly
because he spoke no foreign language and we could only commu-
nicate with him through an interpreter, and partly, I think naturally."
Despite his reticence and the fact that he was burdened with the
job of looking after 1,400 prisoners with only 200 men, the
governor graciously gave them leave to pursue their investigations
in natural history. Yet Nares and Wyville Thomson could see that he
was perplexed. Several times he asked them what port in England
they hailed from and did not seem to be able to comprehend the
purpose of a man o' war adapted for scientific endeavor.

After coffee and cakes, Nares and Wyville Thomson took their
leave of the governor and made their way back to Challenger's
anchorage in the tiny harbor, admiring en route the way that the
convicts caught their fish. They were permitted two pieces of wood
fastened together by cross pieces on which perched a stool. On this
the fisherman sat with a basket and a round coil of fishing line. The
simple catamaran rode so low in the water that, according to Wyville
Thomson, "those fellows . . . look oddly as if they were running
about on the water without any support." The simple construction
of the boats was an effective way of ensuring that no convicts could

leave the island, and even the garrison there did not have anything bigger.

Next morning disaster struck. Overnight, the governor's suspicions had apparently got the better of him and he rescinded his decision to allow the party access to the island. Wyville Thomson was devastated. "We were all extremely anxious to work up this island thoroughly. From its remarkable position nearly under the equator . . . in all its biological relations mainly a South American colony, it presented features of special interest to European naturalists." But the governor was adamant and despite Nares's best diplomatic efforts only a minimum of collecting was allowed.

On the morning of Wednesday, September 3, *Challenger* weighed anchor and left Fernando Noronha. With a heavy heart Wyville Thomson watched the tiny penal settlement disappear over the horizon. "Some of us," he wrote, "had set our hearts upon preparing a monograph of the natural history of the isolated little island." But Campbell, as usual, was more practical and forthright, "I was mighty glad," he wrote, "as it was a stupid little place." Swire at least could understand the factors that had triggered the governor's decision. "It is not much to be wondered at that he should have his doubts concerning us, for we did not carry out the usual routine of saluting the flag, we had been long at sea and consequently were very disreputable in appearance, and lastly the Captain made his visit of ceremony accompanied by his small son Billy and two or three of his savants, which again is not strictly in accordance with ordinary usage on such occasions." And yet, on reflection, Wyville Thomson tended to agree with Campbell's assessment, "I am inclined to think that there was a general feeling of relief on leaving a place which, with all its natural richness and beauty, is simply a prison, the melancholy habitation of irreclaimable criminals."

After the desolation of St. Paul's Rocks and the remoteness and hostile reception at Fernando Noronha, *Challenger's* crew were only too pleased at the prospect of several days rest and recuperation in the mainland port of Bahia on the east coast of Brazil. They arrived there on September 14 after a difficult passage against the southeast trade winds. Finding that they had not been making sufficient way to windward, Nares ordered the sails reefed and steam power used instead. Thereafter they made good time until just outside Bahia, when William Spry rang up from the engine room with the news that they had only two buckets of coal left. The tricky approach to the harbor would now have to be made under sail and the remaining coal kept for maneuvering power as *Challenger* berthed.

Landfall was made successfully, however, amid a cloud of butterflies that enveloped the ship as it maneuvered into the quay. And all agreed that, with one exception, Bahia was a fine place. Campbell summed up everyone's views: "The town of Bahia is from the harbor beautiful; partly built on the face of a high steep bank which, as it recedes on either side of the town, is covered with tropical vegetation, among which palms, bananas, huge aloes and mangoes are visible. Red-roofs, church-spires and domes, yellow walls and coco-palms stand out at the top of the bank; all glitter light and color against a deep blue sky. The town of Bahia from the streets is not so beautiful, and is, to put it mildly, extremely odiferous." Even the equable Moseley, who elected to walk up the steep little streets to the town's center rather than use the sedan chairs favored by the locals, was moved to note, "I ... made my way through steep narrow stinking streets, where the slops were constantly being emptied from upper stories without any warning or 'Gare a l'eau.'"

In all, *Challenger* spent 10 days at Bahia and in that time several expeditions were fielded. Henry Moseley was pleased to be able to naturalize after the thin pickings at St. Paul's Rocks and the disappointments at Fernando Noronha, and immediately made plans for an investigation of the interior. In this he and the others were aided by the generous hospitality afforded them by the local railroad company, which was busy laying track into the interior of the continent. The railroad company, like the local steamship company, was British owned and both enterprises made the weary sailors feel welcome by issuing them with free train tickets. Campbell wrote smugly, "English companies . . . own what ever is enterprising and go-ahead in Bahia," and happily accompanied Moseley into the interior.

They headed about 20 miles inland along a railroad track that was intended to eventually reach the small settlement of Pernambuco. Moseley spent the time observing and making notes, while his companions enjoyed the hospitality of the first-class compartment in which they rode and ate ham and eggs. After some time the train stopped and the companions alighted for a closer exploration of the primeval forest through which they rode, Moseley waxing lyrical about "the immense height of the trees, their close packing and great variety." He noted that the tree trunks were everywhere covered with parasites and climbers such as various species of mistletoe and bromeliads. "The forest was so thick as to be quite gloomy and dark, and as we passed along the path we heard no sound and saw no living animal, except a few butterflies."

That evening they returned to the station, where two cane sofas had been prepared for them to sleep on. The gleam of an oil lamp cast a brilliant pool of light through the open door and they saw that a table had been laid for supper. It was incongruous to sit and dine in such luxury with the thick blackness of the tropical night pressing in just beyond the open doorway, but if anything it heightened their enjoyment. "Then our servant brought in coffee, fried eggs and bread so, with the bottle of wine that we had brought and the game-pate we were not much to be pitied were we?" wrote Campbell. Even Moseley's scientific enthusiasms were not immune to the lure of these creature comforts from home and he noted happily, "Thanks to the energy of the English railway officials, Bass's ale is to be had at all the stations on the line at 2s. 2d a bottle."

Afterward, they strolled out into the darkness, and sat and

smoked while the small boys of the village made "horrible squeal-ing noises by blowing through short conical tubes, made by rolling up strips of palm leaf spirally. . . . Such excruciating sounds seem to be as pleasing to the youthful African ear as to that of the London street boy."

Herbert Swire joined John Murray for a separate expedition into the interior and also enjoyed the extravagances of the British away from home in the heyday of empire. He was particularly taken with the kindness of the railroad company's director. "Mr. Mawson placed a saloon sleeping carriage at the disposal of the officers, to take us whenever we liked, to go in wherever we liked, and to keep up the line as long as we liked." Settling into their personal carriage, they watched as the servant the railroad had allocated them brought on board essential supplies. "Soon afterward we were fairly installed in our new house, with guns etc neatly stowed away overhead, carpet bags etc methodically arranged under the sleeping bunks, and our man making himself generally useful laying in a stock of beer and other necessaries for the journey."

Before they left Brazil, Moseley decided to have one last adventure. A large fair was scheduled 70 miles upriver at Feira St. Anna, a town beyond Caxoeira and, with his interest in local customs at least as strong as his interest in science, he could not pass up the opportunity. Availing himself of the hospitality of the local steamship companies, he headed up river to Caxoeira and then switched to mule. His guide was a German who acted as an inter-preter for the railroad. He was, said Moseley, "a wild sort of young fellow, and had undergone various vicissitudes of fortune," but for all that was "a capital merry companion, knowing everyone on the road and having a joke for all." Moseley enjoyed himself immensely, finding the landscape and the ride on the well-broken mules con-vivial. There was refreshment in abundance, too. "Good Lisbon Wine is sold along the road; the drinking places consist of a hole about a yard square in the gable end of the usual mud-walled cottage, placed at such a height as to be convenient to a man on horseback." After a seven-hour ride they arrived at the tiny town of Feira St. Anna, where they took lodging at the inn, which consisted of a single-story house with a large communal eating room, two communal sleeping rooms, and a kitchen.

It was a perfect opportunity for Moseley to inspect the locals, who, for the most part, were cattle dealers assembled for the next day's market. The beds in the communal bedrooms were packed so closely together that they touched and he was not impressed to find himself sharing a room with a filthy tobacco dealer. And so the night passed uncomfortably, with Moseley rendered sleepless by the fleas, lying in the cramped bedroom, and listening "to the mingled crying of children, barking of dogs, croaking of frogs in the marsh below, and squealing and groaning of the axles of the ox-carts bringing merchandise to the fair."

By 10 o'clock the next morning, the main street of the town was a seething mass of humanity and animals with stalls where farinha, the coarse meal used locally, jerked beef, fruit, and vegetables in abundance were for sale. Other stalls sold needles and thread, still others sweetmeats for the children "but most trying to a naturalists eye," wrote Moseley, "were stalls where various rodents and other small native animals were for sale, spitted on wooden skewers roasted and dried for eating . . . the skulls of all were split open, and they were utterly lost to science." High-quality leather goods were also for sale and here Moseley bought a sturdy leather hat that would last him well for the rest of the voyage, "with this on my head I could butt my way head first into any bush with impunity."

But the most splendid sight of the fair was the cattle market. "The cattle are bred at estates far up the country, where they run wild in the bush and are caught and branded and drafted for market every two years." Moseley quickly developed a deep respect for the vaqueiros, the men who herded the cattle. Dressed completely in leather, they were expert riders and "it is marvelous what work they get out of their small horses." Every now and then one of the cattle broke free and was chased up the street by several vaqueiros. Usually they were brought to heel quickly, but "Sometimes the animals are very fresh and wild, and make off at full pace and cannot be herded. The vaqueiros then strain every effort to come up behind them, catch hold of their tails, and spurring their horses forward so as to get up alongside their beasts, give a sudden violent pull, which twists the animals round, and throws them sprawling on their sides."

There were few breeders at the market, the cattle having been sold by them to local dealers who then re-sold them to the lowland merchants who would take them downriver to Caxoeira or Bahia. The cattle were herded into the pens where the merchants, who never dismounted from the small horses that they themselves rode, could examine them and dicker over the price. Moseley spent a happy day observing the market before spending the night at the house of a friend of his guide. This lodging was on the road to St. Amaro from which Moseley caught a steamer bound for Bahia in the morning.

By noon the next day he was safely back aboard *Challenger*. He found the officers and Scientifics relaxing with new friends. The American corvette *Lancaster* lay in harbor, and "we fraternized greatly with the officers," wrote Herbert Swire, "who turned out to be a capital set of fellows." Indeed, Swire was having an excellent time, he and several of the other officers having struck up great friendships with the members of Bahia Cricket Club with whom they played the sport of gentlemen.

So cordial had relations become, in fact, that the *Challenger* crew had no fewer than three balls to look forward to. But then disaster struck. "During the first few days of our stay a large amount of rain had fallen," wrote William Spry, "this, succeeded by a hot sun and again by rain, formed just the forcing bed for disease." On the day of the ball to be given in their honor by the Bahia Cricket Club, one of the bluejackets went down with Yellow Jack. He was immediately hospitalized and Nares put *Challenger* out to sea, bound under all plain sail for colder lands where the disease could not flourish.

For a time it was touch and go: Would they reach colder climes before the disease had a chance to spread among the lower decks? One man looked as if he had succumbed but speedy action by the surgeon quarantined him on the upper deck, where he was kept under observation. He recovered and before long they were in latitudes where they could feel safe from yellow fever.

Challenger was safe but many were sorry to have said goodbye so abruptly to the friends they had made in Brazil. "On the very day fixed for the ball we left the harbor," wrote Swire, "much to our disappointment and that of our friends on shore. I call it a very delicate attention on the part of the BCC to put our ship's crest at the top of the programmes, and we should very much like to have repaid all their civility by giving some sort of turn-out on board." But it was not to be and by mid-October the ship was within sight of one of its most desolate destinations, the remote, forbidding, island of Tristan da Cunha.

The Library of Time

Tristan da Cunha, South Atlantic Ocean, 37° 03′ S, 12° 18′ W to Simonstown, South Africa, 34° 12′ S, 18° 26′ E

"On the morning of the 15th," wrote William Spry, "land was in sight, a little speck, at first rising up dark and rugged out of the sea, growing larger and larger as we neared, terminating at length in a huge conical peak some 8000 feet in height, covered in snow." It was mid-October of 1873 and *Challenger* had arrived at Tristan da Cunha in the South Atlantic. Some 1,500 miles west of the Cape of Good Hope and more than 2,000 miles from the coast of South America, Tristan da Cunha's claim to fame is its being one of the most isolated islands in the world. It is the largest in a group of five; the others being Nightingale, Middle, Stoltenhoff, and the aptly named Inaccessible Island. Tristan's lonely aspect is compounded by its location in some of the roughest waters of the world, the Roaring Forties, and the island is surrounded by sheer 300- to 600-meter-high cliffs. The weather being fine when *Challenger* arrived, however, they dropped anchor within sight of the only settlement on the island, the tiny hamlet of Edinburgh.

Tristan was discovered in the sixteenth century by the Portuguese navigator Tristão da Cunha but was not colonized until the early years of the nineteenth century when the British set up a colony there to keep a watchful eye on Napoleon, who was incarcerated nearby on St. Helena. To call St. Helena "nearby" is something of a stretch, because it is 1,300 miles to the north. However, the British anxiety to keep "Boney" under lock and key was a reflection of the paranoia that followed his near victory in Europe in the first years of the nineteenth century. In 1817 the British garrison withdrew from St. Helena, but a corporal named William Glass elected to stay there with his family. Over the years the colony was supplemented by other settlers as well as by shipwrecked mariners, and by the time of *Challenger's* visit in 1873 the community comprised about 84 souls distributed among 20 families.

Despite its relatively large size, the island's only habitable portion is a plateau on the northwest shore, where Edinburgh's 15 dwellings were built. The little town had named itself in honor of Prince Alfred's visit six years before. The *Challenger* crew were warmly greeted by the governor of the island, Peter Green, who confirmed that the island community could furnish the ship with provisions. Alfred Taylor and Joe Matkin oversaw the re-provisioning of fresh meat and vegetables, but the warmth of their welcome was somewhat diluted when they discovered that the inhabitants of the island were not, as William Spry put it, "above trying to make a good bargain out of us."

During their visit the crew heard a story concerning two Germans who had arrived at the nearby Inaccessible Island some two years before to hunt for seal fur and oil. Contact with Tristan was established early on but nothing was heard from them for some time, and the islanders, it seemed, were concerned about them. This concern seemed to be typical of the Tristaners and several of the officers noted with approval the islanders' humanity. Wyville Thomson even went so far as to write, "The character of the inhabitants stands deservedly high; they had invariably assisted, to the best of their ability, all shipwrecked persons...."

On October 16, 1873, *Challenger* set sail to the west in search of Inaccessible Island, the crew dredging and sounding as usual and finding that the sea between the two islands was shallow. On Inaccessible they found themselves besieged by penguins that attacked tar, officer, and Scientific alike with sharp-beaked enthusiasm. "The yelling of the birds is overpowering," wrote

Moseley, "I can call it nothing else." To their amazement they also found the two Germans, in good health but without transport. Moseley wrote, "In the early morning, we made out with the glass two men standing on the shore gazing at the ship. . . . They were overjoyed at the chance of escape from the island; we gave them breakfast, and heard their story."

The two brothers Gustav and Frederick Stoltenhoff came to the island in 1871 to hunt seal and penguin but early on lost their boat. This was not of too much concern to them because they had sporadic contact with the initially friendly Tristaners. However, as the months wore on, that friendship degenerated into suspicion and jealousy and eventually the Tristaners sabotaged the Germans' livelihood on the island. By now, the autumn of 1873, the brothers had had enough and wanted passage off the island away from their tormenters.

After hearing how badly the Tristaners had treated the brothers, Nares knew that he had only one choice. He would take the Germans with him to the Cape of Good Hope and make sure they were safely ensconced there before they moved on for the deep south and the Great Ice Barrier. He was not the only one on board to revise his opinion of the good character of the Tristan islanders on hearing this tale, but with no plans to return to Tristan there was nothing more to be said or done.

After a brief visit to the nearby Nightingale Island on October 18, 1873, *Challenger* set sail for Simonstown at the Cape of Good Hope, arriving there on October 28th. There she would stay for six weeks' re-provisioning and refitting for the hardest leg of her long voyage, south to the Great Ice Barrier. The knowledge of the impending journey loomed like a specter at the back of everybody's minds. In part their trepidation came from knowing that only four other expeditions had been to the vast silent fastness of ice at the bottom of the planet since Captain James Cook, one of the most famous navigators of all time, discovered the Antarctic continent exactly 100 years earlier. *Challenger's* voyage would also be groundbreaking, because no steam-powered ship had yet attempted to reach Antarctica. It was this knowledge that hung heavy in the minds of the crew that autumn of 1873. But as if to reassure them, when *Challenger* arrived at Simon's Bay they discovered that she was not the only ship with a fine scientific pedigree there. HMS *Rattlesnake* rode at anchor nearby, carrying recent mail and newspapers from England. *Rattlesnake* had carried Thomas Henry Huxley, one of the *Challenger* expedition's most stalwart supporters, to the South Seas and Australia between 1846 and 1850. Huxley's findings on that voyage formed the basis of his monumental tome, the *Oceanic Hydrozoa*, which, at least in part, explained his support of the *Challenger* expedition.

Rattlesnake had just returned from a skirmish of the Ashanti Wars (or, as Joe Matkin called them, the Ashantee Wars) and her commander, Commodore Commerell, lay badly hurt on shore, seriously wounded by bullets in the lungs and legs. *Challenger's* men had already come across one ship, *Simoom*, bound for this second skirmish between the British Empire and the local Ashanti tribe. It had all been sparked by the local chief's (or *asantehene's*) desire to preserve his empire's last trade outlet to the sea, at the old coastal fort of Elmina, which the British had annexed in 1872. In early 1873, the *asantehene* sent an army of 30,000 warriors to retake Elmina. They were unsuccessful, but managed to gain an important foothold in the area. The British government appointed Garnet Wolseley, a major general in the British Army, as administrator and commander in chief of the region and gave him orders to drive the Ashanti out.

This was the situation when *Challenger* arrived at Simon's Bay in October 1873. The war rumbled on throughout *Challenger's* stay in South Africa until the British eventually won the battle of Amoafo in the late spring of 1874. The Treaty of Fomena then forced the Ashanti to pay an indemnity of 50,000 ounces of gold, renounce their claim to Elmina, end alliances with other warring tribes, withdraw their troops from the coast, keep the trade routes open, and halt the practice of human sacrifice.

To occupy themselves before their Antarctic ordeal, and despite the Ashanti Wars, the officers and Scientifics awarded themselves a holiday. Throughout that Cape summer, as the prevailing south-easterlies blew long and warm from the Indian Ocean, they enjoyed a sybaritic life. William Spry took in the sights. To his eyes, the buildings of Cape Town, some 24 miles from their anchorage off Simonstown, were unimposing and the digs at the local Masonic hotel uninspiring: "the headquarters of Jews, diamond merchants and successful diggers." Indeed all around them was evidence of the new South African diamond rush and more than one of the bluejackets, including the expedition's photographer, deserted, tempted by the lure of an easy fortune. After all, the choice was that or another two years aboard the crowded corvette, and most imme-diately, a long trawl through the worst waters on the planet: the Roaring Forties.

In the 1860s, near the old Dutch colonial town of Zoutkloof, children playing on the banks of the Orange river accidentally dis-covered the diamond fields. The town was located in a region known as Saltpans Drift, which could be reached only by crossing the forbidding Karroo Desert. The children found a large stone that weighed in at a spectacular $21^3/_{16}$ carats. It caught the attention of a trader who sent it to Dr. Atherstone, a local geologist, who pro-nounced it a diamond of the highest quality.

A systematic search soon started and almost immediately turned up a stone of such prodigious dimensions, $83^1/_2$ carats, that it was named the "Star of South Africa." William Spry wrote, "The area over which diamonds have already been found is very extensive, and how much farther it may extend cannot even be conjectured. Sufficient diamondiferous country is already known to provide many years employment for a large population. Diamond-digging is certain to become a permanent industry. . . ." How prophetic!

Today of course, the South African diamond industry is one of the largest in the world

Back at Simonstown the bluejackets' life was quiet and industrious. Joe Matkin, for example, was preoccupied with the task of refitting and re-provisioning the ship for the Antarctic voyage. By the beginning of December *Challenger* was newly rigged, freshly caulked and painted, and fitted with a new deck house for the comfort of the Scientifics. She groaned under the weight of six months' worth of provisions and the stoves now installed in every corner of the ship. Joe Matkin was particularly relieved that one of the stoves was close to his hammock; he, for one, would not suffer through the long Antarctic nights.

Shortly before she left South Africa *Challenger* paid a visit to Cape Town. Two sumptuous balls, on successive nights, were given in the Commercial Buildings there: the first by the *Challenger* men, the second by local dignitaries. "Everything was done by them to ensure success," wrote William Spry, "and, without any flattery, nothing could have exceeded the completeness of the arrangements or the hospitality of the givers." The next day they "swung" the ship in Table Bay harbor to calibrate the compasses before they returned to Simons Bay to complete refitting and loading. At 6:30 in the morning of December 17, 1873 *Challenger* put out to sea.

By the 19th the ship, by now 80 miles south of the Cape, entered the Agulhas Current, a body of warm water that originates in the Indian Ocean and flows initially westward before being forced back in a southeasterly direction after meeting the cold Atlantic easterly drift current. In many ways it is the South Atlantic version of the Gulf Stream and the Scientifics were keen to take its measure. They found that the warm waters of the Agulhas Current extended to a depth of 400 fathoms (more than 2,000 feet) and that the current itself was more than 250 miles wide. Soon after *Challenger* turned her blunt prow south, toward the Marion and Prince Edward Islands, the most westerly of the chain of islands that includes the Crozets and Kerguelen. A few days later and they were among the Roaring Forties where the seas raced unfettered around the planet, and it was here, in this blighted place, that Wyville Thomson finally threw in the towel and capitulated in a long-running argument with his right-hand man about the true nature of the sediments of the ocean floor.

THE LIBRARY OF TIME

The officers, bluejackets, and at least some of the Scientifics had very quickly become bored with the white mud that endlessly filled the dredges, day after day. In his account of the voyage, William Spry refers often to the retrieval of monotonous gray oozes from the seabed. Lord George Campbell, as we have seen, was more forthright in his disgust, and even Henry Moseley, that gifted naturalist, chose to close his own account of the voyage with the words, "as the same tedious animals kept appearing from the depths in all parts of the world, the ardor of the scientific staff even abated somewhat. . . ." In fact, in his excellent account of the voyage, *Notes by a Naturalist*, discussion of deep-sea dredging occupies precisely 2 out of 540 pages! The tedium of dredging and sounding very likely accounted for the high attrition of ship's personnel by desertion.

But two men aboard were not bored by the dredging. Charles Wyville Thomson and John Murray knew very well that achieving an understanding of the nature of the ocean-floor sediments was a critical goal of the expedition. By the time they reached the Caribbean, the chemist Buchanan had already shown that the white Globigerina and Pteropod sediments could be dissolved away to leave the red clay that they had found in the deepest parts of the abyss, but the central, unanswered question was where that white sediment had come from in the first place. Wyville Thomson firmly believed that the oozes were native to the ocean floor. John Murray was equally certain that the oozes were formed at and near the ocean surface and that the white carpets of ooze on the seabed were merely the detritus left behind after shell-secreting animals died and their shells fell through the abyss.

By the time *Challenger* started her transit to the Great Southern Ice Barrier, Murray had gathered abundant evidence that the components of the oozes lived at or near the surface. Indeed, even the scientifically illiterate Lord George Campbell had noticed that, and as they approached St. Paul's Rocks, he wrote "We have found all the globigerinae and foraminiferae, which science (young as yet in these matters) said lived at, and only at, the bottom, alive at small depths below the surface and sometimes on the surface." There could be no doubt that this conclusion was the correct one and Wyville Thomson had to acknowledge that he was wrong and his junior colleague was right.

It was a bitter pill to swallow, because he, together with William Carpenter, his estranged scientific partner back home, had nailed their scientific colors to the mast on this issue. But Wyville Thomson was magnanimous in defeat. By the time they entered the Southern Ocean he had completed a paper for the *Philosophical Proceedings of the Royal Society* on the subject. In it he wrote,

> Since the time of our departure Mr. Murray has been paying the closest attention to the question of the origin of this calcareous formation which is of so great interest and importance on account of its anomalous character and its enormous extension. Very early in the voyage he formed the opinion that all the organisms entering into its composition at the bottom are dead and that all of them live abundantly at the surface and at intermediate depths over the Globigerina ooze area. The ooze being formed by the subsiding of these shells to the bottom after death . . . I have formed and expressed a very strong opinion on the matter. It seemed to me that the evidence was conclusive that the foraminifera which formed the globigerina ooze lived on the bottom and that the occurrence of individuals on the surface was accidental and exceptional; but after going into the thing carefully and considering the mass of evidence which has been accumulated by Mr. Murray I now admit that I was in error and I agree with him that it may be taken as proved that all the materials of such deposits with the exception of course of the remains of animals which we now know to live at the bottom at all depths, and which occur in the deposit as foreign bodies, are derived from the surface.

As *Challenger* traveled around the world, the expedition's artist, J. J. Wild, had plenty time to draw pictures of the foraminifera that are the main component of Globigerina ooze. Many of the expedition report's volumes contain his exquisite drawings of these strange and wonderful creatures. Indeed, one volume, written by the charismatic and ebullient John Murray, is devoted exclusively to the first detailed classification of these sediments. Yet, during and even after *Challenger's* voyage, as Murray was compiling his monumental work, no one could have believed that these tiny creatures would, within a century, become vitally important for deciphering the details of our planet's climatic and oceanographic history. For the geologist and oceanographer there is simply nothing to match the detailed information trapped in the sediment of the deep sea; it is the library of time. Figure 7 shows some of the deep-sea fauna *Challenger* dredged up from the deep.

Today we recognize three types of deep-sea sediments: terrigenous sediments that originate on land, authigenic sediments that were formed in place on the seafloor, and biogenic sediments composed of the skeletal debris of tiny plankton. There are basically two types of biogenic sediment: that formed by the accumulation of siliceous radiolaria and that formed by the accumulation of calcareous foraminifera. The importance of both stems from the fact that radiolaria and planktonic foraminifera live at or near the ocean surface and their shells incorporate a record of surface-water conditions as they grow. But it is the calcareous sediments, formed by the rain of dead planktonic foraminifera through the abyss, that have traditionally formed the backbone of climatic and oceanographic research.

The vast blankets of calcareous sediment formed by these tiny creatures are neither homogenous in composition nor of constant thickness. Calcareous oozes are common in the low and mid-latitudes while in other places, particularly the high latitudes, radiolarian oozes tend to dominate.

The thickness of the calcareous ooze blanket is directly propor-

Benthic foraminifera

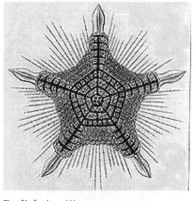

Radiolarian (1)

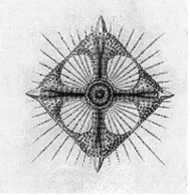

Euplectella subearea

Radiolarian (2)

FIGURE 7 Deep-sea fauna

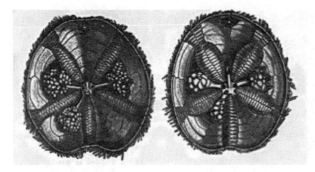

Hemiaster phillipi

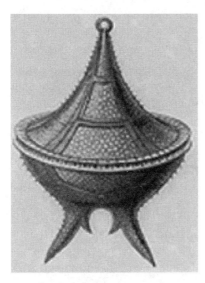

Dinoflagellate (1)

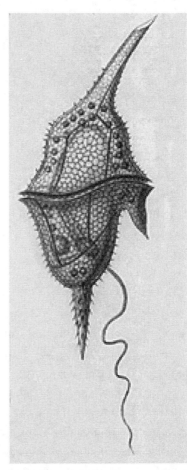

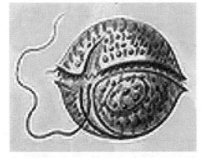

Dinoflagellate (2)

Dinoflagellate (3)

tional to its age. This is simply because new ocean floor, as we have seen, continuously forms at the mid-ocean ridges and then spreads away on either side, aging and collecting a progressively thicker sediment blanket as it does so. It follows, therefore, that the thickest, and oldest, sediments should occur at the edges of the ocean basins and this is indeed what we find. In fact, the oldest sediment occurs at the far western edge of the biggest ocean basin of them all, the Pacific. In this region the deepest sediment is 200 million years old—an age that places it firmly in the middle of the Jurassic period of Earth history. Figure 8 shows the geological timescale.

This overall process varies locally, particularly in areas with a high degree of upwelling that brings nutrients from deep waters and fuels faster than normal surface-water productivity. In such areas, the plankton rain from the surface tends to be enhanced. The most prominent upwelling regions are at the equator, off the west coasts of the continents and off Antarctica. Zones of minimum upwelling and, therefore, productivity, occur in the central regions of the oceans known as the gyres, a good example of which is, as we have seen, the Sargasso Sea. In fact, the accumulation of siliceous sediment is a better indicator than carbonate sediment of high productivity, because in surface waters siliceous organisms tend to dissolve more readily than carbonaceous ones. This means that if siliceous oozes actually manage to accumulate on the seafloor, productivity must have been seriously high. Conversely, in regions below the calcite compensation depth—where the red clays that so fascinated *Challenger*'s Scientifics begin to form—the local calcareous sediment blanket can be thinner because the acidity of the seawater begins to dissolve the foraminifera as they reach the bottom. It was the study of these calcareous sediments that revolutionized the earth sciences in the twentieth century, because it turns out that the forams that make up such a large percentage of the sediment are exquisitely sensitive to climatic change.

The success of the *Challenger* expedition spawned many other cruises, one of the most interesting being the German *Meteor* expe-

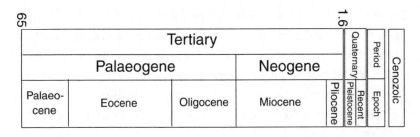

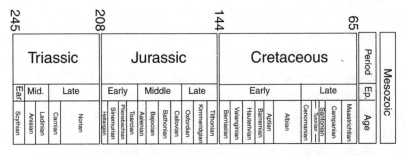

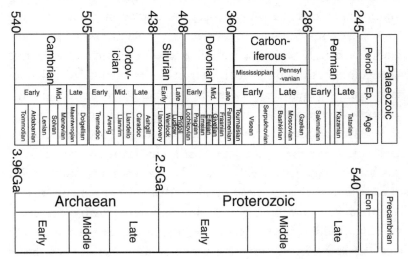

FIGURE 8 Geological timescale

dition of 1925-1927, led by the noted German chemist, Fritz Haber. Haber's claim to notoriety was the development of the mustard gas the German army used in the trenches of the First World War. After the war Haber, aboard the *Meteor*, experimented with extracting gold from seawater. The German government hoped to pay the reparations required by the Treaty of Versailles with that gold. In this the expedition was a failure; there is gold in seawater but the technology of the early twentieth century was not equal to the task of extracting it. However, in other ways the expedition was a notable success. One of the biologists on board, Wolfgang Schott, studied the sediments that *Meteor* dredged up in the South Atlantic.

The technology of piston coring was then in its infancy. The apparatus that Schott used was of the simplest kind, simply a tube with a glass liner and a lead weight to drive it into the sediment. The cores that it retrieved were, therefore, very short, typically less than a meter, but the very simplicity of the apparatus allowed the ordering of the sediments within them to remain almost completely undisturbed. This clever technique enabled Schott to observe that the sediments in all the cores he retrieved were divided sharply into two layers differentiated by the composition of the animals, specifically the planktonic foraminifera, comprising each layer.

The upper layer, about 25 centimeters thick, contained shells of a species of planktonic foraminiferan (known formally as *Globorotalia menardii*) that flourishes in this region today, while in the lower layer this species, and others that were commonly associated with it, were completely absent. Schott made the huge intuitive leap that the lower layer had been formed during the last ice age and the *menardii* and the other species (which together comprise the so-called *menardii* assemblage) were absent because they could not tolerate the lower water temperatures associated with glaciation and so had retreated to lower, warmer, latitudes. It was the first hint that the species composition of deep-sea sediments could be used to infer climate, and it spawned an entire industry devoted to unraveling the mechanisms of climate change.

After the end of the Second World War an ex-paratrooper named Goesta Wollin teamed up with an oceanographer named David Ericson at the Woods Hole Oceanographic Institution (WHOI). Together they took Schott's findings further and revolutionized the study of the ice ages, putting it on a properly scientific footing for the first time. More cores were taken with more advanced coring technology, but everyone agreed that Schott's original finding was correct: The change in the foram assemblage down-core recorded the occurrence of an ice age some time within the relatively recent past. The corollary was obvious: Drill deeper and the changing species richness of the planktonic foraminifera in the sediments might provide a continuous record of climatic history. But first this most recent ice age had to be understood in detail.

In 1948, the steel-hulled ketch *Atlantis* of WHOI anchored over the western terraces of the mid-Atlantic Ridge, giving a new coring device—developed by the father of modern oceanography Maurice "Doc" Ewing—its first trials. One of the big advantages of Ewing's device was that it was simply and ruggedly constructed of stainless-steel tube whose thousand-pound lead weight was extremely effective at driving the corer deeper into soft sediment. The coring tube itself was more than 12 meters long—much longer than that of the expensive and delicate Kullenberg corer used by Schott. On their first attempt they retrieved a core more than 9 meters long, effectively trebling the core-recovery record.

Over the next few years Ericson and Wollin collected cores from all over the Atlantic Ocean as well as from the Caribbean, and corroborated Schott's finding that certain species indicated warm-water (interglacial) conditions while others signified cool-water (glacial) conditions. The warm-water assemblage, as Schott had first observed, was dominated by the foram species *Globorotalia menardii*. The cool-water assemblage was dominated by another species, which rejoiced in the glorious name *Neogloboquadrina pachyderma*.

It appeared that Ericson and Wollin had developed a universal and fantastically simple technique for determining the temperature of the oceans in the past.

On one memorable day in January 1951, they raised a long core from a depth of more than 4 kilometers in the Caribbean region and found that it contained more warm and cold stages than they had ever before discovered. Their excitement was intense. Clearly, this area was uniquely sensitive to glacial-interglacial fluc-tuations in temperature and, importantly, accumulated sediment slowly enough to incorporate more of these variations than usual. By the mid-1950s Ericson and Wollin had compiled data showing that the sequence of the four most recent glacial and interglacial stages was reflected in foram abundances collected all over the Atlantic and Caribbean regions. These results made them confident that their method was the best way of measuring temperatures in the deep ocean.

At this point Ericson and Wollin ran into a conflict with the king of the new science of *isotopic* temperature analysis, Cesare Emiliani. Emiliani had developed a technique first suggested by the Nobel laureate chemist, Harold Urey. Urey's idea was that the proportion of the two common forms (or isotopes) of oxygen in a fossil shell would vary according to the temperature of the water in which it grew. The warmer the water, the higher the proportion of the light isotope of oxygen, oxygen-16; the cooler the water, the higher the proportion of the heavy isotope, oxygen-18. A simple ratio could be converted into an accurate temperature. Urey's idea turned out to be correct and by the mid-1950s the development of oxygen-isotope palaeothermometry was under way in several labs around the world. Because Emiliani specialized in applying the technique to forams, it was inevitable that sooner or late he would run across the work of Ericson and Wollin.

Oxygen-isotope temperature measurements of foraminifera agreed well with the faunal-abundance method of temperature determination near the tops of cores but as one went deeper, significant discrepancies began to appear. By the age of 100,000 years, the agreement between the two techniques was severely compromised. Ericson and Wollin were convinced that their data agreed well with the four continental glaciations known from land-based studies, while Emiliani was equally convinced that his data showed that there were many more than merely four glaciations.

As the analyses continued, the discrepancy got wider and wider and soon Emiliani was claiming that there had been as many as 20 glaciations—or 40 glacial-interglacial cycles during the Pleistocene, broadly the last 2.6 million years. The debate heated up when, in their 1964 book, *The Deep and the Past*, Ericson and Wollin pointed out the difficulties of distinguishing the contributions of temperature and ice-sheet growth and decay to any given oxygen-isotope measurement. The contaminating influence of ice-sheet growth and decay occurs because in times of glaciation the light isotope of oxygen, oxygen-16, is locked up in the ice sheets, leaving the oceans artificially depleted in it. Ericson and Wollin believed that the isotope chemists relied too much on their high technology. They even went so far as to suggest that the high cost of the mass spectrometers needed for isotope measurements led scientists to suspend proper skepticism and believe the numbers too readily.

By 1967 Louis Lidz, a student of Emiliani's, had shown that with some minor statistical massaging Ericson and Wollin's data could fit comfortably after all with the isotopic data. The debate and the bad feeling had been so much hot air, generated by two equally valid techniques that were still in their infancy and whose complexities had not been properly understood. Both the isotope and the foram-abundance techniques for measuring temperature were developed over the next few years.

The problem of how to separate the temperature from the ice-volume effect was addressed by an isotope geochemist at the University of Cambridge in England named Nick Shackleton. Shackleton reasoned that deepwater temperatures would remain stable—at around 4°C—over a glacial-interglacial cycle and that this stability could be measured using bottom-dwelling (benthic) foraminifera. If this deep-water temperature signal was then subtracted from the surface-water temperature signal (from surface-dwelling forams), the difference would be a true temperature signal.

As time went by, however, it became clear that the calculation was not so simple after all. With the advent of the Deep Sea Drilling Project, more and more benthic isotope data from around the world suggested that the temperature of the deep ocean was not as stable as Shackleton supposed. Ironically, help was to come from a development of the foram-abundance temperature technique.

John Imbrie was a gifted mathematical paleontologist who, with his graduate student, Nilva Kipp, had developed a multi-dimensional statistical technique, known as factor analysis, into a way of measuring temperatures over recent glacial-interglacial cycles using subtle changes in the abundance of *several* species of planktonic foraminifera considered, in mathematical terms, simultaneously. When the Imbrie and Kipp technique could successfully reproduce modern-day temperatures from core-top data (that is, recently deposited sediments), they were ready to extend the technique into the past. However, there was one major and apparently insurmountable obstacle: With increasing age, the modern-day species of planktonic foraminifera that the technique was based on become progressively rarer. For practical purposes anything more than about 30,000 years old was beyond the range of their technique. However, in younger strata, the factor-analysis technique proved itself to be in very good agreement with the oxygen-isotope method and gave Shackleton the opportunity to re-estimate the partitioning of oxygen isotopes between temperature and ice volume. Thus began a long-running collaboration between Imbrie and Shackleton, which also spawned one of the most ambitious projects in modern geology: the CLIMAP (Climate: Long-range Investigation, Mapping, and Prediction) project.

The CLIMAP project's goal was to calculate the temperature

of the entire Earth during an average August during the height of the last ice age, 18,000 years ago, using the Imbrie and Kipp and oxygen-isotope techniques. At that time no one was quite sure what caused ice ages, although the suspicion was growing that the theory put forward by the Yugoslav astronomer, Milutin Milankovitch, was correct. Milankovitch suggested that ice ages were caused by cyclic variations in several parameters of the Earth's orbit around the sun: tilt, obliquity, and eccentricity.

The CLIMAP results were spectacular, and in 1976 the team published their results, in the form of a temperature map, in the prestigious journal *Science*. The team did not limit themselves to foraminifera but used other planktonic groups such as the coccoliths and radiolaria. All seemed well, until they started to examine their data from the tropics.

The CLIMAP project showed that the temperature of the tropics had been only about 2°C lower during the height of the last glaciation. This was surprising because the rest of the world had had an average global temperature about 5°C lower than that of the present day. But the transfer functions based on the several different groups examined, plus the oxygen-isotope data, all showed that the glacial tropics had been warmer than expected. However, the CLIMAP data did not agree with data derived from land. In particular, evidence from the fossil distribution of mountain vegetation clearly showed that the tropics had been 5°C colder with a snow-line altitude at least a kilometer lower than today. This discrepancy between the CLIMAP data and the data from the land had profound implications for the new science of climate modeling.

As climate models were developed in the 1970s and 1980s, it became clear that they were very sensitive to the initial parameters that controlled the behavior of the model as the program was run. These parameters are technically known as "boundary conditions." When climate modelers attempt to model future climates, they first calibrate their models by testing them against the past using boundary conditions derived from data from the geological record. When the CLIMAP data set was used to define the critical boundary conditions for the temperature of the tropical ocean at the last glacial maximum, it was found that the models could not successfully predict the temperature of the land estimated from the vegetation and snow-line evidence. The CLIMAP scientists insisted that the problem lay with the accuracy of the models but there was a growing suspicion that there was something wrong with the temperature estimates derived from the CLIMAP data set. In the 25 years since the CLIMAP paper, more and more evidence has been gathered that seems to contradict the project's findings. Data from South America confirm that conditions there were at least 5°C cooler than today, while other results show that the sea surface temperatures near Barbados were also about 5°C colder than today.

Even today the jury is still out on the question of whether the CLIMAP project members got it right. There has been some justified criticism of the transfer-function approach, most notably because it is the rarer species that are the most climatically sensitive. Because the CLIMAP team counted only 300 specimens or so and then multiplied this figure to estimate population abundances, they might have severely underestimated the abundance of important species. However, there is still the fact that oxygen isotopes do agree well with the transfer-function approach. The struggle to reconcile the two approaches and accurately measure ancient sea-surface temperatures continues.

But Charles Wyville Thomson and John Murray were blissfully unaware that their beloved deep ocean sediments would become so important, and cause so many headaches, in the future. It was enough for them to know at last where these sediments came from—the surface of the ocean.

But now all aboard had other preoccupations because they knew that the next leg of their epic journey would be by far the hardest. Ahead of them lay the bleak and little-known islands of Crozet and Kerguelen, inhabited only by penguins and whalers, the whalers often fugitives from justice, and beyond that, the edge of the Victorian scientific world, the Great Ice Barrier of the Antarctic.

The Grim Latitudes

Near Marion Island, The Roaring Forties, 46° 50′ S, 36°75′ E to
The Great Ice Barrier, Antarctica, 66° 40′ S, 78° 22′ E

The voyage from Simonstown had not been without incident. By
December 20 they were well south of the Cape and making a good
10 knots en route to the desolate chain of islands that stretches like
a string of black volcanic pearls across the Southern Ocean, from
Marion Island in the west to the remote and forbidding Heard
Island far to the southeast. It was on that day, too, that the cook and
his mate got themselves literally into hot water. "The cook and his
mate were noticed sitting on the edge of a tub full of boiling water,"
said Campbell, "a heavy lurch came, hot water swashed up and over,
a dismal howl, and well I fancy the cook and his mate will be more
careful in future!"

Dismal was the word for the weather, too, with "squalls, rain
blowing fresh and a heavy swell behind us; ship rolling occasionally
30° each way, air becoming decidedly chillier, chairs dashing head-
long about, breaking their own and everybody else's legs." But they
took some solace from the hot grog that was served on the 21st to
commemorate their first year at sea. In their laboratories, Wyville
Thomson and John Murray labored over their paper on the nature
of the deep-sea sediments and reflected on the successes that the
expedition had amassed in only one year: manganese nodules and
the transects of the Gulf Stream, the Sargasso Sea, and most
especially the mid-Atlantic Plateau. It had been an excellent year
and, as history would prove, of incalculable value to science.

The deteriorating weather had to be combated: "Every petty officer and seaman on board," wrote Swire, "as well as certain of the marines, stokers and others, received *gratis* a thick pea jacket, a large worsted comforter, a pair of mitts for the hands, a knitted jersey, a pair of thick drawers and a sou'wester; he has moreover the choice of either a pair of sea boots or a pair of Flushing trousers, and a blanket is lent to each man, to be returned when called for."

Challenger was now experiencing the "long swell" made famous by Captain Cook a hundred years before. As the corvette was pounded back and forth in seas that raced unhindered around the circumference of the planet, all on board wondered at the tenacity of that tireless navigator and how he had managed to navigate these waters in what was nothing more than, as Swire put it, "a little cockleshell. "

On Christmas Eve, Campbell and some of the others amused themselves by trying to catch an albatross by flinging a baited line over the stern and waiting for the bird to swoop and catch it in its mouth. These were extraordinary creatures, with a wingspan that could reach 17 feet from tip to tip. The capture operation was delicate, because the bird's bill was sharp enough to cut the line if it could get the angle right. This happened several times and the men's efforts were unsuccessful.

The weather was now bitter and their memories of Simonstown and of having to throw water onto the deck hourly to prevent the pitch melting in the heat seemed a distant mockery. It was inconceivable that those memories were only a week old. Now hot cocoa was served regularly on the night watches, while those involved in the sounding and dredging operations received what Swire referred to as a "whack of wine."

At lunchtime, 12 o'clock, on Christmas Day the crew assembled on the deck in the bitter weather and sang "The Roast Beef of Old England." Lunch for the bluejackets was a basic affair, salt pork and pea soup. The officers and Scientifics as usual fared rather better, especially that evening when one of the Scientifics produced a large bun together with several bottles of the Hibernian liqueur known as Mountain Dew. And so Christmas in those grim latitudes passed as pleasantly as possible, given the isolation of their situation. Swire wrote, "I bethought me of what the good people at home would be about at this time, and taking into account the two and a half hours difference in time, I came to the conclusion that whilst I was enjoying my whisky punch they were probably enjoying their dinner, which would be satisfactory for both parties. The toast of "Absent Friends" was drunk with enthusiasm here, as I have no doubt it was at home."

On Boxing Day *Challenger* reached Marion Island, "cold, gloomy looking land, with snow reaching pretty low down," wrote Campbell. Ashore they found the land spongy and boggy with nothing to relieve the eye but acres of moss, fur seals, and seabirds, principally albatross. Here, as elsewhere in the islands of the Southern Ocean, the *Challenger* men failed to distinguish themselves, killing many more animals than was necessary. In their defense we can only remember that they were the product of an age before wildlife conservation. Too, it was the era of descriptive biology par excellence, with the emphasis squarely on collecting specimens and returning them to British museums. Yet the slaughter outstripped even the demands of science and their stomachs. Within minutes of landing at Marion, Swire and the ship's doctor killed a fur seal to no purpose, because they later discovered it was a sea elephant, the skin of which was worthless. Their slaughter of seabirds was even worse. Even Campbell found the harvesting excessive: "Does it not seem a shame to kill these glorious birds for the sake of their wing bones and feet?" he wrote. "Of the first pipe stems are made, and of the second tobacco pouches. . . ." Swire relates how he killed an albatross for a tobacco pouch but made a hash of skinning the feet. The wretched tobacco pouch's value was now zero and the bird had been killed for no reason at all. We may well draw a veil over this more disagreeable aspect of Victorian culture.

The most notable vegetation on the island was the low green shrub, first discovered by Captain Cook, known as "kerguelen cabbage." All agreed that when cooked it was every bit as good the ordinary cabbage they were used to. But by the 27th of December they were ready to leave Marion and, with the weather closing in, it was decided that the proposed visit to Prince Edward Island would be abandoned too. Instead, *Challenger* would head east immediately and seek the Crozet Islands.

Challenger arrived at the first of the Crozets, Hog Island, just before dawn on the last day of 1873. In such high latitudes dawn broke in the wee small hours and by 3:00 A.M. they could see that the island was completely wreathed in fog. All day they waited for the mist to clear, but when the day's end came and it did not, Nares decided to move on for Possession Island, the largest of the Crozet group.

The weather here was better, and by 7:00 in the evening of January 3rd they were in the channel separating Possession and East Islands. It was "a lovely evening," wrote Campbell, "blue-sky golden tinted towards the horizon, and the sun shining brilliantly over the heavy bank of yellow fog out of which we had sailed." Ashore, beneath dark slopes and the black volcanic terraces, they spied a primitive encampment; just a hut, a boat, and some casks. They evidently belonged to sealers but no one was roused even by the firing of the ship's cannon, the sound of which echoed forlornly among the island's narrow valleys. Possession was deserted.

As at Prince Edward Island the unpredictable sub-Antarctic weather frustrated their plans to land. That night a gale blew up, and afterward the same heavy yellow fog came rolling in from the ocean. "We gave up all idea of landing on these abominable Crozets," wrote Campbell, "and made sail for Kerguelen land, running before a strong westerly wind and a heavy swell the whole way."

They arrived at Kerguelen on the morning of January 7, 1874, and could immediately see why Cook had renamed it "Desolation Island" during his visit 98 years before. "Kerguelen land is a gloomy looking land," wrote Campbell, ". . . with its high, black, fringing cliffs, patches of snow on the higher reaches of the dark colored mountains, and a gray sea, fretted with white horses surrounding it."

The island of Kerguelen was discovered by the Frenchman Yves Joseph de Kerguelen-Tremarec in 1772. The news arrived in England in time to reach Cook before he set out on his last voyage, and he was directed by the Admiralty to "proceed in search of some islands said to have been lately discovered by the French in the latitude of 48° south, and in the meridian of Mauritius." Cook found Kerguelen around Christmastime of 1776 and was so overwhelmingly unimpressed with the place that he immediately renamed it. The one redeeming feature of the island was its anchorage, quite the best in the Southern Ocean, which Cook named Christmas Harbor. "The general structure of Kerguelen island," wrote Wyville Thomson, "very much resembles that of the volcanic district of Antrim, or of part of the west coast of Scotland. The coast presents a series of abrupt cliffs and headlands six to eight hundred feet high, terraced with horizontal beds of alternately softer and harder volcanic rocks. Long narrow inlets or fjords, bounded on either side by precipitous cliffs, run far into the land between the ranges of hills, cutting up the island in a singular way into a number of straggling peninsulas connected by narrow necks."

It was indeed a desolate place but the wildlife was as abundant as at the Crozets and on Marion Island. Several parties went ashore exploring that first day and returned with various kinds of seabirds, sea elephants, and penguins. Once again the Kerguelen cabbage was found in abundance. William Spry had already commented on this strange vegetable at Marion Island, where he wrote, "a bountiful providence would seem to have placed it to keep away scurvy from any who might be so unfortunate as to be wrecked on its inhospitable shores."

By the beginning of February they had finished their survey of Kerguelen. It took three weeks to map the island, which was 100 miles long by 50 wide. Yet despite their best efforts, they had not managed to penetrate more than 10 miles toward the center of the island. Joe Matkin wrote,

> The interior of it has never been visited by man, and perhaps never would, for the ground is frightfully irregular and boggy, so impassable all progress is debarred inland . . . the walking was something frightful, the island is one vast swamp. At every other step you sink up to your knees in the boggy ground. What looked like grass from the ship turned out to be moss, and it was the mossy ground which was the most treacherous. Not a tree or shrub was to be seen anywhere, no animals in any sort, neither insects on the earth though we looked carefully, except wild ducks and carrion hawks, we saw no birds, so that we may call it truly a land of desolation.

When the time came to depart nobody was unhappy. The sheer grinding desolation of the place was getting everybody down. Kerguelen presented all over the same dreary and desolate appearance, hills and more hills of volcanic pumice, all covered in snow and with a heavy fog that hung continually over the island so that the interior was always obscured.

However, the island's desolation was offset by whalers who came aboard from the ships *Emma Jane* and *Roswell King*. The *Emma Jane's* skipper dined with the *Challenger* officers almost every night, holding his audience enthralled with bloody tales of their grisly trade in these remote regions. The life endured by the whalers made that of the *Challenger's* men seem a sinecure. The whalers joined up for four years and were paid according to their success, but their average wage was much less than even that of the *Challenger's* bluejackets.

The whalers had been in the area around Kerguelen for three years and in all that time the only inhabited place that they landed on was Tristan da Cunha. Only once a year, when they rendezvoused with a support vessel that brought provisions and relieved them of their cargo of oil and skins, did they see a fresh face. It was an existence that none aboard *Challenger* envied even compared to what they contemplated on the next leg of their own journey, the

300-mile leg due south to the McDonald and Heard Islands and beyond that to the Great Ice Barrier of the Antarctic.

THE MOAT

The seas around Antarctica are the moat that surrounds a fortress, and contribute to that vast continent's isolation. Here in the deep south of the world there are no other continents to break the winds as they blow westward under the influence of Earth's rotation. Consequently, the seas that these winds whip up are some of the most fearsome in the world. This terrible circumpolar storm track and associated westerly current together comprise what is today known as the west wind drift. The west wind drift with its attendant wind and waves, the Roaring Forties, are the barrier that isolates Antarctica from the rest of the world.

Antarctica is the southern edge of all three of the world's major oceans—the Indian, Atlantic, and Pacific. Surface-water currents from these oceans interacting with the west wind drift make the Roaring Forties the world's greatest natural blender, producing a water mass of such chemical and biological distinctiveness that the Southern Ocean qualifies as an ocean in its own right.

Warm surface waters from the tropics of the Atlantic, Pacific, and Indian Oceans move southward along their western margins and are deflected eastward when they encounter the circumpolar current. This interface is called the subtropical convergence and marks the point where the mixing of these different water masses starts. The mixing of warm tropical water with cold surface water from the Antarctic produces an intermediate water type that has its own special name: sub-Antarctic surface water. The Roaring Forties are bounded on their southern margin by another narrow zone where the sub-Antarctic waters of the Roaring Forties meet the truly cold waters that surround continental Antarctica. This is the Antarctic Convergence. But it is the subtropical convergence that is normally taken as the point where the Southern Ocean starts.

The Southern Ocean is characterized by markedly lower temperatures than waters on the northern side of the subtropical convergence, as well as by a much lower salinity, because it contains a substantial contribution of fresh water from melting icebergs. It is this water that, when it sinks, forms Antarctic Bottom Water (AABW) a vitally important component of the deep waters of the world and one that influences the world's weather far beyond the boundaries of Antarctica itself.

AABW owes its unique composition to the fact that Antarctica produces more icebergs than any other continent. Only a minuscule proportion of Antarctic ice melts during the short summer. Virtually all of Antarctica's ice loss is in the form of icebergs, which calve from the massive walls of the Ross and Wurm ice shelves and from isolated glaciers that penetrate to the ocean from the continental interior. Antarctic icebergs carry nutrients, continental sediments, and fresh water into and around the Southern Ocean. As the icebergs melt, this cold, fresh water is added to the maelstrom of the Roaring Forties, which is why the Southern Ocean is, on average, the coldest ocean in the world.

It is ironic that Antarctic icebergs are such an important contributor to the formation of AABW, because the amount of precipitation that falls on the Antarctic continent is, in fact, very small. Indeed the interior of the continent is a cold desert, by far the largest and coldest on Earth. But the continent is so huge, and its storage capacity so voluminous, that even with the lack of precipitation Antarctica's ice fields contain more than 60 percent of the world's supply of fresh water, an amount equivalent to 60 years of global precipitation. Fresh water is Antarctica's most abundant and accessible resource. Each year Antarctica produces some 5,000 icebergs, almost seven times as many as the Arctic and Greenland combined. Antarctic bergs are also much bigger than their northern cousins, each averaging about one million tons of pure fresh water.

Arctic and Antarctic icebergs differ in *shape*, too, a difference noted by Henry Moseley. The *Challenger* crew sighted their first iceberg on February 10, 1874, after weathering a storm of such ferocity that the ship was forced to run under treble-reefed topsails. It was the worst gale that they had encountered since that first Christmas in the Channel. This time *Challenger* was damaged. The seas loosened the weather anchor, which swung free and staved in the hull by the sickbay. The ship's carpenters were up all night repairing the damage; but by the 10th the ship crossed the 60th parallel and the weather lightened considerably.

At first the Scientifics had planned to log each iceberg and calculate a tally that would give some indication of their density in the austral summer. But they had to abandon that plan when they found that more than 40 were in sight at any one time. Moseley was fascinated by the bergs and noted that the typical Antarctic berg was table-shaped, with a thick dusting of snow on the flat upper surface and sides that were perpendicular cliffs. But this basic form extended only as far as the water line, beneath which the bergs spread out laterally, making it hazardous to approach them too closely.

Moseley noted that the above-sea aspect of the bergs showed many variations on the basic tabular form. Partly this was because of the channels around their bases, cut by wave action and contact with the warmer water. These "wash lines" undercut the cliffs at the edges and eventually the ice walls slipped free and crashed into the ocean, often leaving a surprisingly smooth and perpendicular surface behind. As the bergs melted and lost their mass in this way, they tended to roll over, eventually finding a new equilibrium position. When this happened the wash lines lifted away from the sea and ended up at an angle to the water's surface. Moseley realized that it was possible to reconstruct the history of each berg by interpreting the complex interplay between the oldest and youngest wash lines. He noted, too, that even bergs with a complex wash-line history could eventually be categorized because they generally tilted in a regular fashion as their wash lines eroded. To the *Challenger* crew the feeling that Antarctic icebergs were built to a strict design by unseen and little-comprehended forces added to the alien feel of the deep south.

What, then, is the reason for the limited variety of Antarctic icebergs? The difference between Arctic and Antarctic icebergs can be found by comparing the ways they form. Arctic bergs tend to calve from fast-moving glaciers and, therefore, tend to look like small mountains bobbing in the sea. In contrast, Antarctic bergs tend to calve from the more static ice shelves and glaciers that border the continent and protrude into the sea, making them not only larger, but also flatter, resembling the great tablelands of South Africa. Antarctic bergs can also reach immense sizes.

In the 1990s a berg the size of Oxfordshire broke off the Antarctic ice shelf and was widely hailed as proof of global warming. Another, the Trolltunga, began life as a separate ice tongue roughly the size of Belgium before breaking off and floating into the Southern Ocean. Such large bergs cluster near the shore where they chill the already super-cold air flowing from the Antarctic continent and produce huge banks of dense fog. Once beyond the icepack, however, they disintegrate rapidly and most are gone within two months, their melting contributing to the intense cold of the Southern Ocean. The outer boundary of the drifting bergs, where the melting is most intense, is the Antarctic Convergence. The water released by the melting of these bergs sinks rapidly and contributes to AABW.

But the deep waters produced at the Antarctic Convergence are not the only deep waters formed in these grim latitudes. Slightly further north, cold and northward flowing surface waters, sink to become sub-Antarctic *intermediate* water. Both Antarctic deep and intermediate waters spread north of the equator and both exchange with water from the northern hemisphere.

Mixing of water masses in the Southern Ocean is profound and continuous because the waters there are in a permanent state of flux; they never achieve thermal or density equilibrium. A stable surface layer never forms, partly because as icebergs form, salt is concentrated in the seawater, increasing the density of surface water and causing it to sink.

Challenger's Scientifics noted the complexity of the surface and deep-water currents around Antarctica when, on February 11, 1874, at a latitude of 65° S, a detailed analysis of the vertical temperature structure of the ocean showed a layer of deep water (between 300 and 1,800 feet) that was *warmer* than either the overlying or underlying waters. This, almost certainly, was a localized eddy of tropical water trapped between the AABW and the cold surface water formed by the melting icebergs that surrounded the ship. The thermal inversion caused much consternation among the Scientifics, or as William Spry put it, ". . . first a cold stratum to 50 fathoms, and then warmer to 300 fathoms, to the great surprise of our 'Philos,' thus putting their whole theory out of gear. . . ."

The tempestuous whirl of circum-Antarctic waters is also responsible for their being among the most fertile in the world. Upwelling caused by the water racing around the coast of Antarctica brings to the surface nutrients such as nitrates and phosphates, which fuel the extraordinary biological fecundity of the seas there.

Not only does Antarctica transform the waters that circulate around it, but those same waters also transform the continent itself, being largely responsible for the land's extreme cold by maintaining its thermal isolation from the rest of the planet. Water and land are mutually and eternally interdependent and by controlling the formation of the greatest mass of cold deep water on the planet, Antarctica and the Southern Ocean in turn control the world's weather. This phenomenon started when Antarctica separated from South America sometime in the early Cenozoic (or Tertiary era). The single crucial change that ceded control of the planet's weather to this region was its suddenly acquired ability to form cold deep waters, and that was an accident of plate tectonics.

It seems that at the end of the Cretaceous period, just before the dinosaurs and so many other creatures and plants got snuffed, deep waters were formed in the tropics, just about as far as it is possible to get from the poles. The world was still girdled by a super ocean we call Tethys and the planet's surface waters basked in the heat of a sun that was already augmented by greenhouse-gas concentrations several times higher than those of today.

Under these conditions the surface waters of the Tethys started to sink. As they heated up, water evaporated into the atmosphere, concentrating the salt, and the density of the remaining water increased. This deep water then passed out of Tethys into the other oceans of the world and eventually circulated as far as the high northern and southern latitudes. But there is good evidence to suggest that by the end of the Paleocene period—only 10 million years later (55 million years before present)—cold waters were also being formed somewhere in the high southern latitudes. The most likely explanation is that in the Paleocene, when it separated from South America, Antarctica's thermal isolation began.

As Antarctica drifted farther south it became more and more isolated from the rest of the world and the oceans around it widened, initiating the complex system of currents that characterize the Southern Ocean. In a positive feedback loop of progressive cooling, Antarctica became more and more thermally isolated from the rest of the world and cooled still further. The "ice-house Earth" had begun with a vengeance. By the time Antarctica drifted as far as the South Pole, it was the coldest continent on Earth. As early as the Early Eocene the glaciers, which were spreading across the surface of the continent, reached the sea and began cooling the water there. As these waters cooled, they increased in density and began to sink, traveling off into the world ocean carrying oxygenated cold water with them and replacing the warm, saline, deep waters—formed in Tethys—that were previously the dominant type of deep water.

Yet not only is Antarctica heavily responsible for the *control* of global climates through its production of cold deep water, it is also one of the best places in the world to examine the record of past climate, particularly over the last several glacial cycles. These comparatively recent cycles are particularly important to us because they ushered in the era of humankind. And the singular property that makes Antarctica so useful as a monitor of past climatic change is one that Henry Moseley, too, found intriguing. Because by the end of the third week of February 1874, as *Challenger* moved among the icebergs only a handful of miles from the Great Ice Barrier, Moseley observed that the ice of the bergs showed layers of compaction. He wrote, "The entire mass shows a well marked stratification, being composed of alternate layers of white opaque looking, and blue, more compact and transparent ice . . . the color depending on the greater or less number and size of the air-cells in the ice."

It was a clue that in the same way that the sediments of the ocean floor are the library of time, the ice of the Antarctic also preserves a record of climate change.

The Lost World

Great Ice Barrier, Antarctica, 66° 40′ S, 78° 22′ E to
Melbourne, Australia, 37° 45′ S, 144° 58′ E

THE MEMORY OF ICE

On December 16, 1957, the Soviet Union set up the Vostok Ice
Station on the precise position of the geomagnetic South Pole at
the center of the massive East Antarctic ice sheet, as part of their
contribution to the International Geophysical Year. It was an ambitious location for a permanently manned station, because Vostok
routinely records the coldest temperatures on Earth (the coldest
temperature ever recorded, a mind-numbing −89.2°C [−129°F] was
measured there on July 21, 1983). In the early days of the station
there was no air support to lift in supplies but twice a year a great
train of snow wagons made its way up from the coast. By the 1970s,
however, Vostok was being regularly supplied by airlift and it was at
about this time that the station started doing the experiments for
which it is most notable today: taking ice cores to exploit the stratification noticed by Henry Moseley.

The first cores were relatively short, typically less than a
kilometer, and were aimed principally at measuring temperature
variations over the past few glacial-interglacial cycles, if the strata
that represented these cycles could be reached. The Vostok team's
technique for measuring temperatures was a variant of the oxygen-
isotope technique that we encountered earlier, with the important

difference that it was the frozen water of the ice itself that yielded the temperature measurement rather than ocean-dwelling plankton. The Soviets measured the oxygen-isotope composition of the water directly and, like Emiliani before them, were able to recognize the transition from glacial to interglacial climates. However, they were looking not so much at temperature as at changes in ice volume as a measure of glacial-interglacial cyclicity. Remember that as ice sheets grow they incorporate proportionally more of the light isotope of oxygen, oxygen-16, because precipitation originating in the low latitudes is enriched in this isotope. Thus, in the same way that glacial intervals can be recognized in the deep ocean (even in the absence of the temperature effect exploited by Emiliani) by a preponderance of the remaining heavy isotope of oxygen (oxygen-18), glacial intervals can be recognized directly in the ice sheets themselves by a preponderance of oxygen-16. Furthermore, the Vostok team used the enrichment of the heavy stable isotope of hydrogen (deuterium) as a crosscheck. In Antarctica a cooling of one degree Celsius results in a decrease of 9 parts per thousand in the abundance of deuterium, an easily measurable amount.

Using these two techniques, the Vostok team soon discovered that ice representing the last glacial maximum, some 18,000 years before present, occurred at a depth of only about 400 meters. Given that they were retrieving cores at least a kilometer long, this could only mean that the rate at which ice accumulated on Antarctica was very slow compared to other locations, such as Greenland. The prospect of retrieving a complete glacial-interglacial-glacial cycle suddenly looked very good. This was the impetus for even deeper drilling, and by 1998 a core 3,100 meters long had been recovered. It was 50 meters longer than anything recovered from either the Arctic or the Antarctic before, reaching back a staggering 420,000 years and easily covering the last four glacial-interglacial cycles.

But it turned out that much more could be done with the Vostok ice cores. Moseley had noticed that the color of the ice layers varied between white and blue and had understood that the color variations indicated the number and size of air cells in the ice. It was the contents of these air cells that intrigued the Vostok team, because they realized that they contained samples of ancient atmosphere, air samples from the height of the last glaciation as well as from the last interglacial. This approach had recently been pioneered on Greenland ice, but only as far as the end of the last glaciation. The Vostok ice core let scientists examine directly changes in the concentration of atmospheric carbon dioxide, as well as changes in other greenhouse gases such as methane, over several complete glacial-interglacial cycles.

The Vostok team found a close correlation between the concentration of these greenhouse gases and temperature change (measured using oxygen and hydrogen isotopes in the same core), thereby confirming the findings from Greenland that both carbon dioxide and methane concentrations were lower during glaciations than in interglacials. However, the enormously detailed record available from the Vostok core allowed the relative timing of changes in greenhouse-gas concentration and temperature change to be addressed, too. The team found that the transitions from glaciations to interglacials were marked by an increase in greenhouse-gas concentration that *preceded* the increase in temperature, but during the transition from interglacials to glacials change in greenhouse-gas concentration significantly *lagged* behind the onset of cooling. The inescapable conclusion was that changes in greenhouse-gas concentration were important in warming the earth out of glacial periods but the interplay of factors that drove the world *into* glacial phases was more complex.

The Vostok ice core and the other cores drilled in the Antarctic and elsewhere (in Greenland and the Arctic, particularly) eventually yielded much more information than merely temperature and greenhouse-gas change. Variations in the density of windblown dust told how atmospheric wind circulation varied over the past several thousand years, while variations in the concentration of sulfur yielded vital information about the frequency and severity of volcanic eruptions even thousands of miles away. Seventy-three thousand years ago the Indonesian volcano, Toba, erupted with such force that 600 cubic miles of ash were ejected into the atmosphere. The imprint of this event is writ large in the record of the ice sheets, particularly the Greenland ice core known as GISP-2. Based on evidence from this core, it seems the Toba eruption was the most severe of the past half million years. Short-term shifts in climate can be measured in ice cores if the ice accumulation rate is high enough to resolve the events. (The Vostok-core accumulation rate was relatively low, which is why such a long record was retrievable.) Cores from Greenland, again using subtle variations in the abundance of deuterium and oxygen isotopes, show that there was an abrupt global cooling of about 15°C 12,000 years ago that returned the world virtually to the glacial conditions that gripped the world 6,000 years before that. Astonishingly, this climatic shift, known as the Younger Dryas, is estimated to have occurred over only 5 years but then lasted a staggering 1,300 years before the world returned to warmer temperatures.

The imprint of pollution can also be measured in ice cores. Gases trapped in the air cells show clearly an increase in atmospheric CO_2, methane, and nitrous oxides, starting around 1800, about the time the Industrial Revolution got under way. Another, more menacing, type of pollutant that can be measured in the ice cores is radioactivity. In 1986 the Chernobyl reactor exploded, killing 250 people and spewing radioactive fallout across the world. Within only two years researchers coring the Antarctic ice cap had found a record of that dreadful day entombed in the ice as an excess of radioactive elements. Deeper drilling in both the Antarctic and the Arctic clearly show the onset of atomic bomb testing in the 1950s as well as a radiation spike resulting from the frenetic testing that took place in 1965, just before the Test Ban Treaty came into effect. The ice cores of Antarctica show a consistent decline in radioactive fallout since that year.

Finally, one of the more esoteric chemical markers to be found in ice cores is sodium. Sodium is an integral element of that most common of seawater constituents, salt, and it turns out that it can be used as a marker for storminess. In colder periods the latitudinal temperature gradient between the equator and poles is increased and this has the effect of forcing the heat transport engine into overdrive. The heat transport engine is the shorthand name for the process by which the excess warmth of the tropics is transported to the poles either by surface water currents or by air movements. Climatic cooling, whether it is on a short (decadal or century) or a long (millennial) timescale, tends to cool the poles more than the tropics. This forces the heat transport engine to work harder. With the heat engine working harder, the winds and, therefore, the waves, are driven to higher intensity, too, and more of their sodium-laden spume lands on the icecaps. Sodium levels in both the Greenland and the Antarctic ice sheets show that there was a marked increase in storminess around 1400 AD, a time when there was a short period of intense cold that has been nicknamed "the little ice age." At exactly the same time the archaeological record shows that Viking colonies in Greenland disappeared. Could it be that the increased storminess of the little ice age cut the Viking homeland off from its colonies, putting an end to the expansion plans of this ancient seafaring race?

There is much to be gained from the study of ice cores, and the enthusiastic Vostok team drilled deeper and deeper until one of the most extraordinary scientific discoveries of the twentieth century forced them to stop. It was a discovery that would have fascinated the *Challenger*'s Scientifics, too because it was the evidence that they had set out to find in the first place—living fossils.

Termination Land, 61° 18′ S, 94° 47′ E

On February 16, 1874, *Challenger* was as far south as she was destined to go, a latitude of 66° 40′ S and a longitude of 78° 22′ E. She was on the edge of the Great Ice Barrier and only 1,400 miles from the South Pole, the attainment of which would engender so much heartache and privation for the expeditions of Amundsen, Scott, and Shackleton a few decades later. But Nares was under strict orders not to try to proceed any farther. Both victuals and coal were in short supply and the risk of getting trapped in the ice was too great. They would certainly not survive a winter in those grim latitudes. As Wyville Thomson put it, "As the season was advancing, and as there was no special object in our going farther south, a proceeding which would have been attended with great risk to an unprotected ship . . . once or twice the water began to show that 'sludgy' appearance which we know 'sets' so rapidly, converting in a few hours an open pack into a doubtfully penetrable barrier,—Captain Nares decided upon following the edge of the pack to the northeastward, towards the position of Wilkes' 'Termination Land.'"

The expedition's orders made it clear that proving the existence or otherwise of Wilkes's Termination Land was a priority. As recently as 1840 Captain Wilkes of the United States Navy had written,

> In latitude 64°31′ south longitude 93° east, we made what was believed to be land to the south and west, at least as far as terra firma can be distinguished when everything is covered with snow. Soundings were obtained in 320 fathoms, which confirmed all our previous doubts, for on later observation a dark object, resembling a mountain in the distance, was seen, and many other indications presented themselves confirming it. Advancing to the westward, the indications of the approach to land were becoming too plain to admit of a doubt. The constant and increasing noise of the penguins and seals, the dark and discolored aspect of the ocean strongly impressed us with the belief that a positive result would arise in the event of a possibility to advance a few miles farther to the southward.

In fact, the idea of a continent in these southern regions was a lot older than Captain Wilkes. As early as the beginning of the eighteenth century, a continent had been marked on early charts, in the complete absence of any hard geographical data, and given the name "Terra Australis Incognita." Its existence was hypothesized because the geographers of the day felt that something was needed to counterbalance the weight of the continents of the high northern latitudes around Greenland and the Arctic.

By the 23rd of February *Challenger* was approaching the area marked on the charts as being where Wilkes had spied his great continent, but there was nothing to be seen. The weather was fair, with a slight breeze from the northwest that fell away to nothing by noon. All hands strained their eyes for some glimpse of land but to no avail. It seemed inconceivable that such a landmass should be invisible at such close proximity. There were icebergs aplenty, however, as well as strange cloud formations. Joe Matkin wrote, "On several occasions land had been reported from our mast head, and had it not been for our having steam, we might have marked new lands down on the Southern Charts; but although we could all have sworn that we saw land on one occasion, on steaming towards it, it proved to be a peculiar and remarkably defined vapor cloud. We have passed over several spots marked on the Chart as "Indications of land" without finding any and I daresay a steamship would dispel several of the mythical discoveries in this part of the world."

By evening they had steamed right up to the edge of the pack but still there was no sign of land. The view, however, was beautiful. Wyville Thomson wrote, "The ship and the ice were for a time bathed in an intense yellow light, which faded into a delicate mauve, with cold patches of apple-green between the clouds. A long roll of heavy cloud stretched across the sunset sky, and the golden glow which it took after the sun went down was truly magnificent."

During the night the weather changed. By daylight on the morning of the 24th the wind was rising fast; the sky was dark and threatening, with frequent snow squalls that blew across the ship blinding clouds of wheel-like crystals so intensely cold that they burned when they touched the skin. At 4 A.M., they deployed the dredge in the hope of getting something before the weather got too bad but had to pull it up before it passed 1,300 fathoms. It was time to seek shelter and this Nares did by bringing *Challenger* into the lee of a large iceberg. The idea looked good until, during a sudden lull in the weather, *Challenger* lunged forward and ran into the giant berg, smashing the jib boom and dolphin striker and carrying away several other items of the head gear. By now the sea was running hard, Force 10 on the Beaufort scale, estimated Wyville Thomson, and the euphoria brought on by the beauty of the ice pack was being rapidly replaced by terror at the thought of an icy lonely death.

It was indeed the coldest day that *Challenger* had yet experienced and in the violent seas she lay to under bare poles, just trying to survive. By lunchtime fog had descended and visibility was reduced to less than 50 yards. At 3 o'clock that afternoon it lifted just enough for them to see a large berg drifting directly toward them. The ensuing confusion, wrote Joe Matkin, "was something fearful; nearly everyone was on deck, it was snowing and blowing hard all the time; one officer was yelling out one order, and another something else. The engines were steaming full speed astern, and by hoisting the topsail, the ship shot past it in safety."

The storm raged all night and three lookouts were posted. Nine times that night *Challenger* put about to avoid collision, and by the morning of the 25th the ship was covered in half an inch of snow. Despite the weather, the ship's two carpenters were busy and by the 26th they were able to replace the jib boom. But the weather was still dirty and Nares himself spent that entire night out on the deck seeking shelter among the bergs when he could, or running from them when prudence dictated. "It was," wrote Matkin, "considered the worst and most dangerous night we have had. . . . Altho' we were all eager to see an iceberg, we are just as anxious to lose them now, it is so dangerous sailing these foggy nights with such masses of destruction all round us."

There was still no sign of Termination Land so the hunt for it was abandoned. Spry wrote, "Wilkes' vision was at fault, and the great Antarctic continent has turned out to be a Cape Flyaway. . . . Having now proceeded as far south as practicable in an undefended ship, at noon course was altered to the east. . . . On reaching clear water studding sails were set and we were off for Australia, Cape Otway, 2, 278 miles distant."

So *Challenger* turned its back on the Antarctic, much to the relief of all aboard, from Captain Nares and Charles Wyville Thomson down to ship's steward's assistant Joseph Matkin. But all aboard would have been aghast to know the scientific discovery that they were turning their back on, a discovery that would be made a hundred years later by the Vostok Ice Core team.

THE LOST WORLD

In the mid 1970s—almost exactly a hundred years after *Challenger* was in the region—an airborne radar survey of the ice sheet around Vostok Station revealed the presence of an enormous freshwater lake, beneath the ice sheet, that covered an area of about 10,000 square kilometers. By the early 1990s observations from satellites confirmed the presence of the lake, which was named Lake Vostok, and the news was released to the public by the Russian and American scientists who made the discovery.

The news caused a sensation, because the central question was obvious to all: How could a lake of liquid water survive under 4 kilometers of ice in the coldest region on Earth? The answer is simple: The ice sheet is heated from below by heat flowing outward from the interior of the Earth, known to geophysicists as the geothermal flux. Add to this the fact that 4 kilometers of ice is an excellent insulator and the presence of liquid water becomes understandable. The phenomenon is much the same as that which allows pools of water to accumulate below the slow-moving glaciers of the Swiss Alps, but on a vastly expanded scale in the Antarctic.

In the mid 1990s the Vostock scientific team had put together all the data that had been collected over the past 30 or so years and added new findings to provide a fuller description of Lake Vostok, and their interpretation was extraordinary. The lake was not only huge, it was in places extremely deep, as much as half a kilometer. Although early estimates had suggested that the water in the lake might be quite old—at least 50,000 years—as knowledge of the lake grew, it seemed likely that the deeper parts were cut off from the usual circulation pathways (by which water arrives by melting of the surrounding ice sheet and leaves by sliding out from under the ice at the lowest point of the lake) and might be very much older, perhaps as much as *one hundred million* years. The implications of this finding were staggering, because it meant that it might be possible to sample liquid water that was contemporary with the dinosaurs. Before drilling was stopped to preserve this pristine, ancient environment, microbes—bacteria and fungal cells—were found in the lowermost layers of the ice cores. Today it seems likely that Lake Vostok *is* a repository for ancient life, a true lost world of which Conan Doyle's Professor Challenger (probably another namesake because Conan Doyle studied under Wyville Thomson in Edinburgh) would have been proud, and which the Scientifics aboard HMS *Challenger* would have given their eye teeth to find.

This is the irony, of course, for as we have seen, one of the main reasons for the voyage of HMS *Challenger* was the desire to test the Darwinian notion that the oceans would be the repository for life forms that were, from the evidence of the fossil record, thought to be extinct. It turned out, however, that the ocean bottom was not so much the lost world that they sought, but rather a lake under an ice cap. However, it was a lake that was as inaccessible to them as the surface of another planet. So *Challenger* got within a thousand miles of Lake Vostok and the answer they sought before turning obliviously north for Melbourne and warmer climes.

Today one wonders whether any of the Scientifics could have had any inkling of what they might be missing by turning their backs on Antarctica. Almost certainly not. But the presence of these ancient bacteria and fungi has implications beyond merely those of evolutionary curiosity, because with modern techniques of DNA sequencing (in conjunction with the so called polymerase chain reaction) we really can now assess the genetic structure of life forms thought to be long dead. This has huge implications for the pharmaceutical industry because it offers the prospect of a completely new set of templates on which to base drugs. The waters of Lake Vostok could be the greatest natural pharmacy in the world.

But there is more, because it turns out that Lake Vostok is not unique. Since its discovery back in the 1970s at least 70 other sub-Antarctic lakes have been discovered, most of them in the same area of the East Antarctic ice sheet. But if Lake Vostok is not unique, there is still one place for which it and its companions might well be a unique *analogue*, and that is Europa, one of the moons of Jupiter. Europa, like Antarctica, is covered by a thick sheet of ice, only in Europa's case it is frozen methane, not frozen water. Soundings from spacecraft show that there is liquid water under this unimaginably cold crust, which immediately leads us to ask: If there is life under the Antarctic ice sheet, could there also be life under the Europan ice crust? The National Aeronautics and Space Administration (NASA) certainly takes the possibility seriously, and that is why drilling has been stopped at Lake Vostok. The lake provides a perfect opportunity to test the techniques that will be needed to sample alien life forms without contaminating either their environment or ours. The scientific community is now hard at work devising robots that can be rigorously sterilized before being sent down to sample the ancient waters of Lake Vostok. They will be the progenitors of the unmanned probes that we send to search for alien life.

In the late 1970s the first indications of Europa's true nature began to emerge during the flyby of *Voyager II*, the second of the pair of space probes sent to explore the outer planets and their satellites. The Jupiter system is so far from the sun that our star is merely a flicker in the distance, a firefly glow that gives no warmth.

The *Voyager* missions were, for their time, every bit as ambitious as the voyage of HMS *Challenger*, and it is ironic that they were approved by the U.S. Congress at about the same time that Lake Vostock was discovered, a hundred years after *Challenger* started out on its own epic journey.

The impetus for the *Voyager* mission was to exploit a rare configuration of four of the outer planets, Jupiter, Saturn, Uranus, and Neptune, which occurs only every 175 years. This configuration meant that a spacecraft could visit each of the four planets using the "gravity-assist" technique perfected on an earlier *Mariner* mission to Mercury to slingshot its way between them. This eliminated the need for vast quantities of propellant as well as large on-board propulsion systems. It also meant that the journey time to Neptune could be reduced from 30 years to just 12. Although the "four-planets" mission was known to be possible it was judged that the likelihood of a spacecraft lasting the distance was so low that the mission was initially limited to visiting just the nearer two, Jupiter and Saturn.

Indeed, the likelihood of failure was judged to be so great that two identical spacecraft were built and named *Voyagers I* and *II*. So far from the sun would the *Voyager* spacecraft venture that each was equipped with a controversial nuclear power supply. They were launched in the late summer of 1977 from Cape Kennedy in Florida. *Voyager II* left first, in late August, on a slower trajectory that would allow it continue with the four-planets mission if by some miracle it survived. *Voyager I* left in early September on a shorter, faster trajectory that would send it north out of the Solar System after its encounter with Saturn.

Voyager I arrived in the Jupiter system in March 1979 and almost immediately made a spectacular discovery. Io, one of Jupiter's moons, showed evidence of localized but extreme volcanism. This was astonishing, because volcanism showed that this moon was at least locally warm with temperatures of up to 15°C, perfectly warm enough for life.

Voyager II was due to encounter Europa within months. Scientists wondered if Europa too would provide surprises like those at Io. It was already known from long-range, ground-based spectroscopic studies that the surface of Europa was covered by a thick layer of ice. If there was volcanism on Io, could there not be volcanism on Europa, too? And if there was, could there not be life there as well?

In those days all deep-space probe missions were run from the California Institute of Technology's Jet Propulsion Laboratory (JPL) in a massive, amphitheater-like room not unlike Mission Control in Houston, which oversaw the manned *Apollo* missions to the moon.

On the evening of the Europa flyby several luminaries, including the sage of the search for extraterrestrial intelligence himself, Carl Sagan (with a group of his students), assembled to watch on the overhead monitors. The telemetry was primitive and as they waited, the group speculated about what they might find and how they would know whether there were volcanoes under the Europan ice sheet. The suspicion was that some hint of these buried volcanoes would be visible as smooth patches where the ice had locally melted and erased any hint of craters left by ancient meteorite bombardments. Smooth patches would indicate recent volcanic activity. But they were not ready for what they saw when the first image grainily established itself on the screens.

As the image slowly coalesced they saw, not a partially pock-marked ball of ice indicating localized volcanism and ancient meteor bombardment, but a perfectly smooth sphere of ice with no evidence of meteor bombardment *at all*. The conclusion was inescapable and its implication undeniable: The surface of Europa was being continuously cooked by global under-ice volcanism, and there was a massive planet-girdling water ocean underneath the ice cap. Europa was suddenly the best candidate for extraterrestrial life anywhere in the Solar System.

Amazingly it was to be 18 years before the JPL returned to Europa for another look with better equipment. Part of the delay was because of pressure on NASA and JPL resources to operate other missions and part was due to the catastrophic explosion of the space shuttle *Challenger* in 1986, which effectively shut down America's manned and unmanned space program for two years. (The *Challenger*—Orbiter Vehicle-99—was *also* named in honor of

HMS *Challenger*, as was the final lunar module capsule to land on the Moon during *Apollo 17*.) But eventually new missions were planned and launched. In 1997 the *Galileo* spacecraft reached Europa, equipped with the latest in imaging equipment and the very first thing that it showed was the presence of icebergs on Europa. This could mean only that in places the ice was being warmed enough to break the shiny smooth shell that encircled the globe and allow chunks of ice to tilt. It meant, too, that the water ocean that had to underlie the ice sheet on Europa must be relatively close to the surface, perhaps as shallow as a mile. The Europan ice sheet was probably thinner than that of Antarctica.

Today the consensus is that there is an ocean of water, probably salt water, underneath the Europan methane ice sheet. The question of whether there is life in that ocean must remain open until we can send a probe to land on Europa. Objections have been raised that so far from the sun there is not enough free energy to support the type of life based on photosynthesis with which we are familiar. However, this might not be a problem, because the same energy that maintains water in a liquid state on Europa, the massive gravitational tidal forces of Jupiter and its satellites Ganymede and Io, could support life similar to that found near deep-ocean hydrothermal vents on Earth. Another suggestion is that life could tap the energy of the charged subatomic particles that swirl through the Jovian satellite system powered by Jupiter's magnetic field.

The next step is to send a probe to land on Europa and sample it directly for life. However, the same concerns about contamination that stopped the drilling just above Lake Vostok apply, perhaps even more directly, to the idea of exploring Europa. At present JPL is developing a robot probe to explore both Lake Vostok and Europa. The Lo'ihi Underwater Volcanic Vent Mission Probe is currently being tested in an investigation of an undersea volcano 20 miles southeast of Hawaii's Big Island at a depth of more than a kilometer. The probe was specially constructed to withstand the most stringent sterilization procedures, and operate in extremes of temperature and pressure.

As I write this, the *Voyager* spacecraft are still functioning after more than 25 years in space, four times longer than their design lifetime. They are now the most distant human-made objects in the galaxy. *Voyager I* visited Saturn and then, constrained by its more limited trajectory, started to leave the plane of the ecliptic heading north. *Voyager II* was able to exploit its slingshot orbit and visit both Uranus and Neptune, thereby fulfilling the original four-planet mission envisaged by the mission designers all those years ago. It is now heading out of the Solar System to the south of the ecliptic. Both spacecraft are well on their way out to the heliopause, the place where our Sun's influence finishes and the galactic wind of interstellar space starts. Like the voyage of HMS *Challenger* itself, theirs is an epic voyage of scientific discovery; but unlike *Challenger*, the *Voyager* spacecraft will not be returning home.

We salute them.

Arrival. Melbourne, Australia, 37° 45′ S, 144° 58′ E

It took *Challenger* 16 days to make the passage from 61° 18′ S, 94° 47′ E (the position of Wilkes's mythical Termination Land) to Melbourne. After the rigors of their Antarctic sojourn those aboard might have been forgiven for hoping for an easy run, but it was not to be.

On March 2, 1874, they almost ran into a large berg during the middle watch. The frequency of bergs had declined so much that they were no longer running under reefed sails and were bowling along at a brisk pace. Swire wrote,

> The practice of shortening sail had been abandoned, and so we stood on all night when, suddenly in the middle watch the lee lookout man came aft and confidently told the officer of the watch that he "thought as 'owe there was an 'hiceberg about one hundred yards right ahead." The ship was going eight knots and had not the officer of the watch seen the berg before the lookout man the chances are that *Challenger* would have

ended her career that night. How many ships that have "not since been heard of" owe their destruction to careless look outs in ice-berg regions!

On top of this, food was running short. They were down to their last sheep, and after that they would be out of fresh meat. Swire had mixed feelings about the fate of the poor animal. "We have only got one live sheep left and after he is killed have to trust to preserved grub and ship's provisions, which I much deplore. I fancy the feelings of the last sheep out of the forty that we took on board at the Cape; he who wrote about the "Last Man," could perhaps likewise depict the harried sensibilities of this, our last mutton. I really think that after surviving the thirty-nine chances against him he is deserving of his life. But we are selfish where it concerns our interior economy, so I fear old fleecy-back must die."

By March 15th they were approaching Melbourne, Australia. The sense of anticipation aboard the ship was palpable and all hands applied themselves to the task of making *Challenger* look presentable. It was also the day that they made contact once again with the human race. As Swire put it, "Today has been a beautifully fine, dry day and this afternoon a phenomenon appeared on the horizon in the shape of a full, rigged ship, the first we have seen for many a weary day. Preparations for Melbourne are in full swing on board, now; paint and whitewash are being slapped about in the most promiscuous manner, so that one is half poisoned by the smell of turpentine etc. I shall try to get in shore as soon as ever we drop anchor."

On March 17, 1874, *Challenger* dropped anchor off the leafy suburb of Melbourne known as Sandridge. Except for the exploration of Marion Island and Kerguelen, they had been three months at sea without setting foot on land.

The Echoes of Evolution

Melbourne, Victoria, Australia, March 17, 1874, 37° 45′ S, 144° 58′ E to
Sydney, New South Wales, Australia, June 8, 1874, 33° 55′ S, 151° 10′ E

It had been the longest period of the voyage away from civilization
and the crew and Scientifics shared a sense of relief when the
Antarctic leg of the voyage was finally over. No one complained
when Nares scrapped the planned visit to Hobart in Tasmania and
decided instead to head straight for Melbourne. Almost as soon as
they anchored, *Challenger* was overrun with influential citizens
anxious to see the ship and to bid its crew welcome to Australian
waters. As they rode at anchor in Hobson's Bay they were amazed
and delighted by the contrast to the silence and loneliness of their
Antarctic sojourn. All was movement, color, and gaiety. Coastal
steamers arrived and departed, yachts and steam launches filled with
pleasure seekers and cargo crisscrossed the bay, and all around them
merchant ships and other men o' war were towed in and out of the
harbor as they arrived from distant lands or made ready to set out
across the Pacific Ocean that lay just beyond the sandbars.

Most impressive of all, the horizon before them was dominated
by the vast, magnificent city of Melbourne, its skyline broken here
and there by church spires and high roofs and the foreground dominated by the new buildings of government: the city hall, the new
post office, government house, the parliament buildings and the
treasury. Everything they saw reminded *Challenger's* crew that they
had arrived at the furthest outpost of an empire that ruled the

world, and that Australia, like India, was another vibrant jewel in Victoria's crown.

All aboard, especially Joe Matkin, eagerly awaited news from home. Matkin was not disappointed, receiving seven letters in all, including one from his mother with news of the wedding of his beloved brother, Charlie. Yet not all his news from home was good. Joe's father was still unwell so the young man lost no time in writing his beloved papa a letter enclosing a payment against the loan that Mr. Matkin had made him before he left Portsmouth 18 months and many thousands of miles before.

But even this news could not dull Joe's enthusiasm for Melbourne. He had been to Australia twice before and both times had thoroughly enjoyed himself. The city of Melbourne was three miles distant from the port but easily accessible by train, for which the authorities furnished them free passes. Matkin and the others wasted no time in leaving their cramped ship and heading off to explore the city. Victorian Melbourne was the commercial capital of Australia and although it had been settled only 40 years before, by the time *Challenger* reached the city it was already thriving. Indeed it was officially classed as the ninth largest city in the British Empire. Those *Challenger* men who had not yet been out to this far-flung corner of empire had heard stories of the legendary Australian hospitality and they were not disappointed. Spry reflected that the months of March, April, and May 1874 were among the happiest of his naval career and it was no surprise that when the ship weighed anchor for the short voyage north to Sydney at the start of April, many of the crew deserted to stay on in Victoria.

Challenger arrived in Sydney on Easter Monday, April 7, 1874. Among those unfamiliar with the "Australian station," the approach caused some consternation. The harbor was almost completely enclosed by high cliffs and inexperienced hands wondered if they were about to run aground. But almost at the last minute the gap between Sydney Heads became visible—as Captain Nares knew it would—and *Challenger* slipped through into one of the most per-

fect natural harbors in the world. They anchored between HMS
Dido, fresh out from Britain, and the German frigate *Ancona*. As at
Melbourne, the men were much impressed by the fine array of
modern buildings that surrounded the harbor and particularly by
the Astronomical Observatory, with its tower and time-ball, which
dominated the grassy slopes on the cliffs above. Although Sydney
Harbor was arguably one of the most spectacular and beautiful
anchorages in the British Empire, like any working port it had its
more prosaic side and the eastern shore was dominated by wharfs
and quays where freight and passenger vessels berthed. Here, too,
was the dry dock *Challenger* desperately needed after the rigors of
the Antarctic voyage. She was still scarred after her encounters with
icebergs so proper repairs to her jury-rigged jib boom were a top
priority.

The months in Sydney passed agreeably, the officers at least
finding the company civilized and charming. Below decks, however, the bluejackets agreed that they preferred the more earthy
charms of Melbourne. The cultural pretensions of Sydney reminded
them too much of the stratified class system that prevailed at home
and aboard *Challenger*.

On the evening of June 7, 1874, the sea was calm, with not a
breath of wind disturbing the tranquility of Sydney harbor. William
Spry stared out across the water as the lights began to appear in the
city. He was in a pensive mood and found himself remembering
with amazement that it was only 102 years since Captain Cook
landed at Botany Bay, just a few miles away over the ridge in front
of him, and claimed Australia for the empire. How far the citizens
of this far land had come in so short a time! As he watched the
lights' reflections heave in the oily swell he reflected on how pleasant
the last three months in Australia had been and on the price those
aboard *Challenger* paid for the chance to see the empire in this way.
Tomorrow they left for New Zealand, ordered by the Admiralty to
complete the final leg of surveying needed to trace the route for
the new telegraph cable between Sydney and Wellington. Although

they planned one more stop in Australia, at the tip of the far northern promontory known as Cape York, the vastness of the Pacific
already beckoned and with it the certainty of more long months at
sea in the cramped confines of the ship. That was the price, Spry
knew, the men paid for the chance to see the empire.

Below decks in his airless cabin, Joe Matkin's thoughts were
similar as he put the finishing touches to letters that he would have
to take immediately to the post office. Almost two dozen men had
gone missing in Australia, tempted by the extraordinary beauty of
the land, the opportunities for settlement, and their distaste for
another two years of cramped confinement in the ship. And yet Joe
stayed on board: As he stared at the soot-scarred bulkhead, he could
not for the life of him imagine why.

Henry Moseley was as sorry as the others to be turning his
back on this wonderful continent, but he was moved not so much
by trepidation for the voyage ahead as by regret for leaving a place
with such singular zoology. It had become clear to him that if the
deep ocean was one place where the proof of Mr. Darwin's theories
might be found, then Australia was another.

CHAINS OF COMMAND

The Australian writings of both Spry and Moseley contain frequent
references to the so-called missing links of the fossil record,
reflecting Victorian biologists' preoccupation with the search for
intermediate forms of life, that is, forms that have characteristics of
both ancestor and descendent. This is another powerful testament
to the influence that Darwin's *Origin of Species* had on the formula-
tion and execution of the *Challenger* expedition. Just as Darwin had
asserted that forms of life that were found on land only as fossils
would be found alive in the depths of the ocean, so had he set the
cat among the paleontological pigeons by stating that the rarity of
intermediate life forms in the rocks of the geological column was
due to great gaps in the fossil record. These missing links. he thought,

were due to subsequent earth movements, times of zero sedimentation, or because evolution had proceeded in isolated enclaves that left no trace of themselves in the fossil record.

Such missing links were inferred from the way that Darwin himself saw evolution proceeding: by the long slow accumulation of favorable variation over the course of eons. A prerequisite for Darwin's theory was vast spans of time, geological time, over which these favorable variations could accumulate and thereby form new species. Darwin did not believe that evolution could proceed rapidly. In fact he explicitly denied the possibility in *Origin of Species*. It is difficult, with our twenty-first century perspective, to see why Darwin eschewed rapid evolution, but perhaps it was because of his innate Victorian conservatism. He had already served up a large lump of heresy by suggesting that animals and plants evolved from other animals and plants in the first place, and perhaps he worried that if he suggested that the process occurred quickly he would fatally damage the credibility of his theories. On the eve of *Origins* publication, Thomas Henry Huxley himself, Darwin's most pugnacious defendant and, as we have seen, one of the main proponents of the *Challenger* expedition, had cautioned the sage of Down House against this view in a letter. "You have loaded yourself with an unnecessary difficulty in adopting *Natura non facit saltum* so unreservedly" (the Latin tag means "Nature does not make leaps"). But the prevailing wisdom was that progress—including evolution— was slow and measured and that missing links, whether in the fossil record or squirreled away in obscure corners of the empire, were there to be found if one looked hard enough. It is no surprise, therefore, that when the *Challenger* expedition set out, only 13 years after publication of *Origin*, the Scientifics had the finding of these missing links high on their agenda.

One of the most spectacular of the missing links was found by an expedition that Wyville Thomson led into the wilds of Queensland in the last days before *Challenger* departed for New Zealand. Spry reports that the professor returned from his foray "laden with botanical wonders and fishing spoils, (among which were several specimens of a peculiar fish called "*Ceratodus*" or, popularly, "Barramundi")."

Ceratodus (or *Neoceratodus* as it now more formally known) is the Australian lungfish, a member of an obscure group of fish known from other continents, particularly South Africa and South America, that have lungs to supplement their gills as well as rudimentary limbs for walking upon the land. Technically, they are a member of the *Dipnoi* (the lungfish), which are themselves part of a larger group called the *Sarcopterygii* (the lobe-finned fish). The importance of the *Sarcopterygii* is that it is the group from which land-dwelling vertebrates (tetrapods) evolved about 300 million years ago during the Devonian period. Walt Disney celebrated the importance of this seminal evolutionary transition in his groundbreaking animated film *Fantasia*. Perhaps you remember the sequence wherein a curiously shaped fish with blobby fleshy fins struggles to make it out of a shrinking pool of water and onto the land under the light of a broiling sun? Well, that was a sarcopterygian, a lobe-finned fish.

Today the lungfish, such as *Neoceratodus*, are the only surviving members of the lobe-fins. Even in *Challenger's* day the Scientifics were very well aware of the colossal importance of the vertebrates' transition from sea to land. They knew that their find of *Ceratodus* was hugely important. In fact, the fish had been described only a few years earlier, in 1870, by the German naturalist Johann Krefft. However, Krefft's choice of publication to describe his new fish must have made *Challenger's* Scientifics wince. Conventionally, new species are named in the pages of a learned scientific journal and it is inconceivable that Krefft, as curator of the Australian Museum in Sydney between 1864 and 1876, was unaware of this. Yet he chose to publish his finding in the *Sydney Morning Herald*! Significantly, Krefft mistakenly described the fish as an amphibian, a characteristic that reinforces how central the lungfish is to understanding the ancestry and evolutionary relationships of the tetrapods.

There are only three genera of lungfish alive today and each is unique to a single continent. The South American version is called *Lepidosiren*; the one from Africa *Protopterus*. Both have twinned lungs and well developed fleshy fins that they use for scrabbling about on the banks of their native freshwater pools and rivers. But the Australian type appears the most primitive, with poorly developed fins and only a single lung sac. The Australian version is very similar to a fossil lungfish known from the Mesozoic era, also named *Ceratodus*. Because of this use of the same name, and particularly because there is no fossil record linking the two, the Australian lungfish has, since the time of the *Challenger* expedition, been renamed *Neoceratodus* (that is, new *Ceratodus*).

There appear to have been at least seven families of lungfish widely distributed around the world in the Paleozoic era, quite unlike their very restricted toeholds today. But after the big extinction at the Permo-Triassic boundary (the boundary between the Paleozoic and Mesozoic eras, see Figure 8) this number dwindled to two (including the long-ranging and widely distributed *Ceratodus*). Today the lungfish are restricted still further, to only three genera.

Even today the precise ancestry of the tetrapods remains unclear. The current best estimate is that they evolved from a lobe-finned ancestor also shared by the lungfish and the coelacanth, that strange fish with large fleshy fins discovered off the coast of South Africa in the 1930s and widely hailed as a living fossil. Whatever the lungfish's evolutionary affinities, though, it is clear that the first fish to crawl out of the sea and onto the land all those millions of years ago during Devonian time cannot have looked very different from the sad relics left behind today. But there is another ancestor of ours even closer to home and it, too, attracted the attention of *Challenger*'s Scientifics, particularly ship's naturalist Henry Moseley, in that spring of 1874.

That spring of 1874 Moseley went out on one of three missions from Sydney for the purpose of collecting specimens. He was particularly excited about the possibility of capturing a duck-billed platypus, *Ornithorhynchus paradoxus*, which were allegedly common on the banks of the Yarra River. The platypus and its relative, the spiny anteater (the echidna), excited his curiosity because they were the only two species left of the egg-laying mammals, the monotremes. To distinguish them from the marsupials (metatheria) and the placental mammals (eutheria) the monotremes are placed in their own class, the prototheria.

Mammals are characterized by the ability to suckle young with milk produced from special glands (the mammaries). Both the platypus and the echidna can do this, too, though their milk-producing organs are more primitive than those of higher classes of mammals (the marsupials and placentals). Like these higher mammals, the platypus and the echidna also have a jaw composed of a single bone, three inner-ear bones, relatively high metabolic rates, and hair. So apart from their curious ability to lay eggs, their membership in the mammals is well warranted. However, it is this egg-laying ability that demonstrates the monotremes' close relation to the reptiles, in particular the group of reptiles known as Therapsids or mammal-like reptiles.

The Therapsids thrived in the latest Paleozoic and the early part of the Mesozoic (the Permian and Triassic periods) and gave rise to several lines of descendents including one, the cynodonts, that we believe is the branch that leads directly to the mammals. But the monotremes probably split away from this main line of evolution early on, in the late Jurassic or early Cretaceous. Modern monotremes retain many of the features of their Therapsid ancestors. For example, the girdle of bones that support the shoulders—the pectoral girdle—is complexly constructed and there is only a single common opening, the cloaca, for the elimination of body wastes and for reproductive matters. The cloaca is a distinctly reptilian feature and is quite unlike the separate openings typical of marsupials and placentals. The skulls of monotremes are almost birdlike in appearance and it is well known that birds evolved from reptiles;

indeed there are some who say that the dinosaurs never really became extinct at all but are with us still in the form of the birds.

The monotremes' curious combination of very primitive and very advanced features (for example both the platypus and echidna use advanced electrical-current-sensing apparatus to track prey) is an excellent example of what has come to be known as mosaic evolution, where different structures within the body evolve at different rates. We know today that mosaic evolution is powerful support for the Darwinian theory of evolution by natural selection. It shows that evolution acts upon structures that must be upgraded to allow species to successfully colonize different niches while leaving organs that are irrelevant to this process unchanged. In the same way that the lungfish is a missing link between the proper fish and the tetrapods, the monotremes are missing links between the reptiles and the mammals.

However, it is important to recognize that neither is a living fossil. Both have been highly modified by subsequent evolution. Their retention of these primitive characteristics suggests that evolution has not acted to change them as it has with certain other features. The lungfish and the monotremes are missing links in only a limited sense: They are a *distorted* reflection of the true, now long extinct, missing links between fish and tetrapods, and between reptiles and true mammals.

However, before we leave the monotremes, it is worth noting that Moseley and the other Scientifics must have been well aware of their importance. Moseley reports disappointment that he was not able to examine the eyes of the platypus "to see if the retina contains brightly pigmented bodies, as in the case of reptiles and birds...." In all the time he was in Australia he was destined never to see a live platypus or even a kangaroo. The only platypi he came across were delivered to him by the Aboriginals and they decomposed in the heat before he could get them back to his dissecting bench aboard *Challenger*.

The voyage from Sydney to Wellington was fraught with difficulty. Their first attempts to leave Sydney Heads on June 8th foundered in bad weather and despite an emotional farewell from their many Australian friends as well as the other two British ships in port—*Dido* and *Pearl*—on June 10th *Challenger* was forced to make a humiliating about face and return to Sydney. On June 12, 1874 though, she finally left Australia.

Despite the continuing bad weather, for the next two weeks they dredged and sounded, trying to find the best route for an undersea telegraph cable from Sydney to Wellington. About 200 miles from the Australian coast they found that the bottom shoaled to a depth of less than 1,700 feet and became exceedingly rocky. This meant that when it came time to lay the cable a more robust one would be needed. The crew congratulated themselves; this would be important news for the cable engineering teams.

By the time they sighted Cape Farewell in New Zealand, the weather had worsened and *Challenger* was forced to take shelter for a night and a day in Port Hardy, an inlet in the northern end of D'Urville Island in Cook Strait. The next day they managed only another short run of some 20 miles before being forced to shelter in the lee of Queen Charlotte Sound's Long Island. The following day they made it across Cook Strait, *Challenger* creaking and groaning in the high sea. When the towering pinnacles of Ben Mor and the Karakora Ranges loomed up across the spume, all aboard felt relief.

They were only 10 miles from harbor when tragedy struck. With the heavy sea still running and shallow water under the ship's keel, Captain Nares set a young seaman named Edward Winton to sounding. At 12 o'clock, lunch was piped and all but a skeleton crew went below for their midday victuals, leaving Winton perched precariously out on the narrow platform known as the fore-chain.

It was then that Winton's sounding cable caught round the anchor. One of the marines still on duty reported later that in a frantic effort to free the fouled line, young Winton—25 years old and married only days before he left England—climbed up the anchor chain. At that moment a heavy wave struck the ship, smashing plates in the mess and shaking the ship from stem to stern, causing much hilarity in the mess; but up on deck poor Winton had vanished. It was almost 10 minutes later that the marine noticed the vacant fore-chain. Nares immediately telegraphed the engine room to stop and had the ship put about and set off back over its track. It was to no avail. Winton had been wearing a complete set of thick winter woolens, oilskins, and sea boots, and despite being a strong swimmer had not stood a chance in the heavy sea. After steaming back and forth for an hour the search was abandoned.

Just before sunset *Challenger*, with a stunned and saddened crew, entered the great sea lake known as Port Nicholson. They made anchor among the other ships at the Queen's Wharf in Wellington. Joe Matkin wrote, "The passage is over at last and so is the long voyage for one poor fellow . . . drowned in broad daylight whilst performing his duty, and not one of us to see him go or to throw him a life buoy; the ship went steaming on for Wellington and he to his last home."

The next day both officers and ordinary seamen sent the hat round for Winton's widow and raised £50. Eventually the memory of the nightmare passage from Sydney began to fade, but if those on board needed any reminder of the treachery of these waters, they got it only a week later when news reached them that the *British Admiral*, an iron steamer on its maiden voyage out of Liverpool, had struck a reef near Melbourne and went down like a stone, taking with her nearly all of her 80 passengers and crew.

Challenger did not linger in Wellington. The Admiralty wanted her back in Spithead by the spring of 1876 and still she had the vast bulk of the Pacific Ocean ahead of her. But *Challenger's* encounters with missing links were not yet over, because Moseley found, among a pile of rotting wood in the forest outside Wellington, one of the most extraordinary examples in the animal kingdom.

THE INSECT WORM

Peripatus is one of those creatures that really do make you believe in evolution, because it is an animal with the body of an earthworm but the appendages of an insect. It is the missing link between the Annelids and the Arthropods. There are about 70 known species of *Peripatus* from New Zealand, South Africa, South and Central America as well as Australia. The group is normally confined to the tropics but some species are found in the temperate latitudes of the southern hemisphere. There are no northern-hemisphere, temperate-dwelling species.

The formal name for the group to which the species of *Peripatus* belong is the *Onycophora*, derived from the Greek word for claws (onychos) and bearer (phoros). The animals range from 1.5 to 15 centimeters long, have between 14 and 43 pairs of walking legs, and exhibit a wide range of color. The name "velvet-worm" is often used for *Peripatus* because of its soft, flexible cuticle. This soft cuticle allows it to burrow into dark moist crevices and protect itself from the water-loss to which it is especially vulnerable, and its hydrostatic skeleton, which is common to all annelids, supports and maintains its shape. The hydrostatic skeleton is a fluid-filled cavity in the body wall, supported by a layer of smooth muscle, which holds the animal rigid in much the same way the air inside a balloon holds it rigid. Having this skeleton is a major clue that the onycophorans are related to other annelids such as the common or garden earthworm. The other similarity is in the way that *Peripatus* excretes waste from its body: through a pair of hair-lined ducts called nephridia.

Like other annelids, *Peripatus* has a truly segmented body (so called metameric segmentation) that confers multiple redundancy on certain body components such as muscle systems. Metameric

segmentation is found in all animals more advanced than the annelids but might be so modified as to be almost unrecognizable externally, as is the case with arthropods. Metameric segmentation of the
body into functional units should not be confused with superficial
segmentation (for example of legs or other appendages) that is
commonly seen in animals such as the arthropods. The legs of
Peripatus are quite unique and do not show the arthropod design.

The features that do link the *Onycophora* to the arthropods are
the haemocoel, a blood-filled cavity in the center of the body that
serves as the circulatory system for distributing oxygen and
nutrients; legs; blood sinuses divided lengthways along the body; a
heart structure similar to other arthropods; and jaws derived from
modified walking legs.

The *Onychophora* have a long and distinguished fossil record
and indeed are known from that most celebrated of all fossil
localities: the Burgess Shale of British Columbia. The rocks in this
locality are slightly more than 500 million years old and are a snap-
shot of the time soon after the Cambrian radiation. The Cambrian
radiation is the name given to the explosive increase in the diversity
of animals and body plans that occurred soon after the Precambrian-
Cambrian boundary (the end of the Proterozoic eon) 540 million
years ago. A feature of the Cambrian radiation was the evolution of
fossils with hard limey skeletons, which greatly increased their
chances of preservation in the fossil record. As a result, the Cambrian
radiation marks the beginning of the Phanerozoic eon, the age of
visible life.

The Burgess Shale fauna was discovered in 1907 by Charles
Walcott, geologist and secretary of the Smithsonian Museum, while
on vacation with his family in the Canadian Rockies. One day,
while riding along a little-used path, he discovered several blocks of
limestone with exquisitely preserved invertebrate fossils in them.
He eventually traced their origin to a rock layer high above the
path and over the next several summers excavated a quarry there
that is still known as Walcott's Quarry.

The Burgess shale fauna lived on the seafloor in the shadow of a massive carbonate reef, which itself sat just off the western margin of the ancient continent of Laurentia (part of present-day North America). Periodically the reef shed large quantities of mud in the form of slides and it was one of these that entombed the fauna now found in Walcott's Quarry.

The range of fossils found in the Burgess Shale is extraordinary and many exhibit body plans that have no living counterpart today. It appears that the early Cambrian was a time when evolution experimented with several different body designs and discarded many. Many that did survive are now but a vestige of their former abundance, represented by only a few species. One such group is the *Onycophora*. All the various species of *Peripatus* alive today are grouped in a single genus and all live on land. However, the Burgess Shale boasts two marine species of onycophoran, the common *Aysheaia* and the extraordinary *Hallucigenia*.

Hallucigenia caused much controversy when maverick Cambridge paleontologist Simon Conway Morris first described it in 1977. As you can see in Figure 9, the creature was portrayed as standing on two rows of inflexible spines, its back bearing a row of tentacles that were thought to be some form of feeding apparatus.

It was one of the organisms that defined the Burgess Shale's reputation for weirdness and was thought to have no living relative—one of the Cambrian's failed experiments in evolution. A decade later *Hallucigenia* was compared with a new fossil from a Cambrian site in China. The two looked very similar except that the Chinese fossil had the twin row of spines along the back and the mobile appendages were found to be paired. These were interpreted as legs. Re-examination of Conway Morris's type specimen (that is, the single fossil declared to be representative of the species as a whole) showed that there was evidence for another row of flexible appendages. They, too, were paired. *Hallucigenia* had been reconstructed upside down and when it was turned on its head it bore a remarkable resemblance to other fossil onycophorans as well

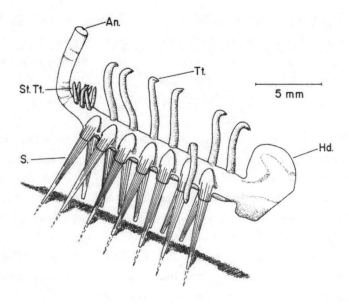

FIGURE 9 *Hallucigenia* as originally constructed by Simon Conway Morris. From S.J. Gould, *Wonderful Life*.

as *Peripatus* itself. The only unique thing about *Hallucigenia* were the two rows of stiff spines along the back, which were probably a defense against predators. To put it loosely, *Hallucigenia* was nothing but an armored *Peripatus*.

Today, although the onycophorans are merely a remnant of their former diversity, they show several fascinating features. Because they forage mostly at night, they evolved only primitive eyes and use stubby antennae to feel their way about. They have another set of modified limbs, called oral papillae, which secrete a sticky substance that dries on contact with the air but that they can squirt as far as 50 centimeters. They use this to deter predators as well as to capture their prey, insects and other small creatures. The third set of modified limbs on their heads are jaws. With these they break open the cuticle of their prey and inject a corrosive saliva. They then ingest the partially digested flesh through a sucker-shaped mouth.

Little is known about onycophoran reproduction except that the male packages his sperm into a bundle and attaches it to the female's cuticle. The female's white blood cells break down the body wall beneath these spermatophores, allowing the sperm to enter her body, where they find their way to her ovaries. Some onycophorans lay the fertilized eggs outside the body but most retain them inside. Certain species even nurture their growing young with a form of placenta, a common strategy in mammals, but rare in invertebrates. The young of these placental onycophorans leave the mother's body as fully formed juveniles.

Onycophorans, like other arthropods, have a system of narrow pipes or trachea that branch inward from the surface of the body for gas exchange. Above the pharynx they have a well-developed brain that links with the rest of the central nervous system via a pair of nerve cords. These nerve cords reunite near the rear of the animal to form another small brain, an anatomical feature considered primitive among invertebrates.

Although onycophorans have a mixture of traits that effectively make them a cross between annelid worms and the arthropods, they, like the monotremes and the lungfish, are highly modified missing links and, like the monotreme, and the lungfish too, diverged from the main evolutionary line (in this case the line leading to the arthropods) early on.

Challenger's visit to Australia and New Zealand was an eye opener for the Scientifics. The region was without question rich in the evidence that Mr. Darwin so longed for, intermediate forms between major animal groups. The Antipodes were indeed lands where missing links were no longer missing. In this place the *Challenger* expedition had found a large part of what it sought, the very echoes of evolution.

Challenger stayed less than a week in New Zealand and forewent the planned trip to Auckland. The weather was foul and Campbell was singularly unimpressed by this depressing outpost of empire. "At Wellington we found the governor staying, so instead of remaining only a couple of days, and then going on to Auckland, we stayed the whole prescribed New Zealand time there, where there was nothing to be seen and less to be done. . . ." The weather was wet and windy and nobody was sorry when they set sail again, on July 7th, for Fiji and the Friendly Isles. Before they left New Zealand, though, Campbell noted one peculiarity of New Zealand house construction. "Earthquakes necessitate building of houses out of wood, slight shocks frightening Wellington occasionally; one in particular 26 years ago partially shook the town down, thereby causing panic."

On July 10, 1874, *Challenger* crossed the International Date Line for the first time, arriving in Tonga on July 19. It was a Sunday and the natives, stark naked, flocked out to meet the ship, paddling their outrigger canoes with enthusiasm. But they would not trade on the Sabbath, preferring instead to attend no fewer than five church services. John Wesley's missionaries had been active on the island for 50 years and the men of *Challenger* were gratified to observe this successful importation of western culture.

On the Monday *Challenger* took on board supplies of fresh fruit and meat and then departed the Friendly Isles, heading for Kandavu in the Fiji Islands. En route they spied flying fish of the genus *Exocetus* as well as humpbacked whales, and enjoyed some of the most spectacular sunrises and sunsets of their entire voyage. At Levuka in the Fiji Islands, Captain Nares granted shore leave to those who wished to go and many of the ratings became involved with the local rum, a noxious brew of such potency that several had to be hoisted back on board like cattle. One young man who had come aboard at Sydney to help make up the deficit left by deserters, drank so much that he became quite mad with the alcohol and had to be put in chains. He was left just outside Joe Matkin's door. All night long he raved but the next morning was in a pitiful state, with a black eye that he could not remember acquiring, apologizing to all and sundry and foreswearing drink for evermore. Matkin, though, was not convinced by the young man's alleged repentance, and noted with some irony, "It is easy to see that his years are numbered for this world."

On Sunday, August 30, 1874, *Challenger* returned to Australia, passing through the Great Barrier Reef not far from the low-lying, three-quarter-mile-long hummock known as Raine Island. She was now in the Coral Sea, a body of water that less than a hundred years later would become infamous as one of the bloodiest battlegrounds of the Second World War. In May 1942, the Japanese Fleet under the command of Admiral Inouye was pressing hard to gain a foothold in northern Australia. But after the debacle at Pearl Harbor the Americans were now ready to put up a fight. Admiral Nimitz directed the carriers *Yorktown* and *Lexington* into the Coral Sea to defend not only northern Australia but also Port Moresby in New Guinea. Throughout the first week of May the battle raged and ended with an American victory. It was an important point of the war in the Pacific and thereafter the Japanese were forced, little by little, back to their home islands. To this day the Coral Sea is littered with the wrecks of the warships sunk there. To scuba enthusiasts it makes some of the best diving in the world.

But even in *Challenger's* time the Coral Sea was hazardous because of the reef density. She stayed there for only a week at the end of August 1874 before turning west and traveling the last 120 miles that would take her to the most northerly point of Australia, Cape York. She arrived there on September 1, 1874, and anchored off the small community of Somerset. Joe Matkin was unimpressed with the area, describing it as having a "flat sterile appearance" with "not a hill over 600 feet high." What did impress him, however, was the contrast between the flat monotony of the northern Australian scrubland and the huge mountains, some more than 13,000 feet high, that loomed only 120 miles away, across the Torres Straits, on the island of Papua New Guinea. Not only did those mountains fascinate him, but he knew exactly what they were, describing them as part of a

> [M]ountain system [that] may be traced across the Pacific from the Rocky Mountains in North America. It rises in the Sandwich Islands over 13,000 feet, again in Japan higher still, still higher in New Guinea. It does not touch the continent of Australia but stretches more to the east; it appears moderately high at the New Hebrides and Fiji Islands and extends through the New Zealand islands rising over 13,000 feet on the South Island. . . . Everywhere along its course earthquakes are prevalent and owning to its near vicinity they have occasional shocks here that are felt in no other part of Australia.

Joe Matkin knew what he was looking at all right: one of the biggest and most extraordinary marvels that *Challenger* would encounter in all her epic voyage, the most westerly point of the ring of volcanoes that rim the Pacific Ocean, the so-called Ring of Fire.

The Groaning Planet

Cape York, Australia, September 8, 1874, 10° 56′ S, 142° 40′ E to
Yokohama, Japan, June 16, 1875, 35° 28′ N, 139° 38′ E

THE RING OF FIRE

By the early 1960s Bruce Heezen and Marie Tharp's seafloor maps
were revolutionizing geology. The idea of seafloor spreading was
catching hold in geology departments around the world and geologists were excited about it. Everybody knew that the old order was
changing. The new theory of seafloor spreading was, to the earth
sciences, every bit as significant as Darwin's theory of evolution had
been to the biological sciences a hundred years before. But one
central question remained unanswered: If new seafloor was being
produced at the mid-ocean ridges what was happening to planet
Earth? Was it simply getting larger like an inflating balloon, or was
that seafloor eventually being consumed somewhere so that the
status quo could be maintained?

It was at this point that Heezen's celebrated geo-instinct failed
him, because he chose the expansion explanation. To accommodate
the new crust being formed at his seafloor spreading centers, the
Earth was simply increasing in size. It was Harry Hess, the chairman
of the Princeton geology department, who put the final pieces of
the puzzle together correctly. Like Heezen and Ewing, Hess was
first and foremost a synthesist as well as being a geologist of wide
experience. He solved the spheroidal, three-dimensional jigsaw of

the theory that would come to be known as plate tectonics by announcing the existence of subduction zones, regions of the Earth's crust where material formed at the mid-ocean ridges is returned to the mantle.

He published his theory in a paper entitled "History of Ocean Basins" in 1962. Crucial to Hess's synthesis was the memory of a submarine trip he had made as a graduate student to the Puerto Rico Trench (the deepest part of the Atlantic Ocean) in 1932. He was accompanying the great Dutch geophysicist, Felix Andries Vening Meinesz, on a mission to measure local variations in the strength of gravity. They discovered that the gravitational field at the bottom of the Puerto Rico Trench (and other trenches, as it turned out) was strangely weak. Meinesz suggested that this could be because low-density, continental crust was under these areas, being forced down into the interior of the Earth and displacing the higher-density material of the mantle that should have caused higher gravity in these areas.

In the white heat of the continental drift revolution of the late 1950s and early 1960s, it occurred to Hess that these trenches at the edges of the oceans were perfect places for the crust being formed at the mid-ocean ridges to be returned to the Earth's interior. This recycling phenomenon was eventually called "subduction" and the trenches named "subduction zones." For Hess, planet Earth was like a caldron of boiling water bubbling at the center (the mid-ocean ridges) and then convecting sideways before cooling and falling back into the interior of the pot (the marginal trenches). Quite suddenly, Wegener's theory of continental drift had the mechanism that it had previously lacked. It was now clear that the continents were carried around the world on the back of the blocks of crust.

As Joe Matkin stared across the Torres Straits at the mountains of New Guinea rising into the haze, he had no idea that he was looking at one of the consequences of these subduction zones, which is that they are almost always accompanied by mountains, specifically, volcanic mountains. This is because as the old oceanic

crust is forced underneath continents—as is happening at the western edge of the Pacific Ocean—water trapped in the rock is released at depths of about 100 kilometers in the mantle. This water,
together with gases such as carbon dioxide, accelerates the partial
melting of the already-hot crust. The melted rock, being less dense
than the surrounding mantle material, starts to rise, forcing its way
back up into the crust by locally melting the rocks above it until it
bursts out on the surface of the Earth in the form of a volcano
chain above the subducting crust.

The formation of volcano chains above subduction zones is
almost universal on our planet but the ocean floor is not always
subducted simply beneath the edges of continents. In some places
ocean floor sinks beneath *adjacent* areas of ocean floor and the
volcano chains form islands. *Challenger* would arrive at one of the
greatest examples of this just a few months later.

TWILIGHT OF AN EMPIRE

On September 8, 1874, *Challenger* weighed anchor and moved
gently out into the Torres Straits. Campbell was not sorry to be
leaving the blighted territory of the Cape York peninsula and the
difficult circumstances of their anchorage there. "Once more in
Australia!" he wrote, "and a horrid country it would be, if it were all
like Cape York. The anchorage is in a narrow straight separating a
small island from the mainland. Through this straight tides run with
the greatest swiftness, necessitating our dropping two anchors, and
wearing one's life out with anxiety during the night watches."

They were heading north into the Arafura Sea, a broad shallow
pocket of water bounded by New Guinea to the east and Timor to
the west. Beyond Timor another arc of volcanic islands, barely
glimpsed from their current position, curved gracefully northwest.
This chain incorporates Sunda, Sumatra, and Java. Between Sumatra
and Java, in that year of 1874, lay an island that within a decade
would become infamous across the world by blowing its top with a

vengeance and causing spectacular sunsets for years to come: Krakatoa.

Yet as the Australian coast receded into the southern haze, *Challenger* sailed toward the vestiges of one of the greatest of the European empires, the Dutch East Indies. The heart of this trading enterprise was a group of islands known as the Moluccas. Even in *Challenger's* time they were famous across the world as the Spice Islands because of the variety and quality of the spices to be found there. The ship approached Banda, the heart of the Spice Islands, having lingered in the Aru and Ki islands off the west coast of New Guinea, toward the end of September 1874.

Challenger was now in the heart of the Dutch East Indies and for days all aboard had felt the presence of a colonial power that was not their own. It was even more discomforting to know that the heart of that power lay just a handful of miles from England across their own Channel and still more discomforting to be reminded how distant—how very distant—that narrow, precious strip of water was from them now. In this far-flung outpost of the world it was a reminder that the British were not alone in their imperial ambition. Joe Matkin wrote "Our men o'war seldom go this way . . . this route to China is called the Molucca passage and is little frequented by any but Dutch ships, so here they reign supreme and have their own way." The weather, though, was clement, "calm as a mill pond," and a welcome relief after the tribulations of the passage to New Zealand.

Banda Island lies just to the south of the island of Ceram (Seram as it was in those days) and was well liked by *Challenger's* men despite its use by the Dutch colonial administration as a convict settlement, for it was beautiful. They anchored on September 30, among wooded islands of vivid green. Their anchorage was landlocked between three islands, two tiny and one a "grand volcano cone" said Campbell. The Dutch colonial houses were graceful and large, set back amid cocas, kanary, and nutmeg trees, while the rest of the small town lay half hidden in the foliage of palms. But it was the volcano that impressed them most, rising 2,200 feet above the tranquil scene, the vast secret bulk of the "ever burning" volcano, Gounong-Api.

The cone was a reminder, if one were needed, that they were still in a land of tectonic unease, "People here expect small earthquakes continually and severe ones occasionally, during which last they are not surprised if all the ships in the harbor are chucked up on to the land, and if all the houses come tumbling down," wrote Campbell.

The Dutch might have been imperial rivals but the aristocratic young officer was charmed by them. He loved the brightly attired people and the trim, whitewashed houses, each surrounded by its own garden where, in the shade of evening, Dutch families assembled. But he saw too that these same houses were surrounded by high walls, all heavily buttressed against the regular earthquakes.

Challenger departed Banda and its vast nutmeg plantations on October 2, arriving in the Spice Island capital, Amboyna, two days later. There they stayed until the 10th, when they moved on for the island of Ternate to the north. On either side, striding to the north were lines of volcanic sentinels, grim reminders of the fiery temperament of this part of the world.

They made their way through the Molucca passage, past the islands of Obi and Batchian where the clove trees grew in glorious abundance, then past Makian, the old volcano that had exploded in 1646, splitting the peak in two and destroying villages wholesale. On the 15th they passed between the great symmetrical volcanoes that dominated the islands of Ternate and Tidore, and anchored off the town of Ternate in the evening.

Despite the looming presence of the gently smoking peak of Ternate, Campbell's enthusiasm for yet another earthly paradise fairly bubbled over.

> Cloves, pepper, cinnamon, nutmeg, coffee, cocoa, pineapples, durians, oranges, limes, citrons, bananas, bread-fruit, and endless others, with palms of every kind, are all here planted on a greensward level as a

billiard table. I never saw in any country anything more delicious and
beautiful, and the only plantations which can be at all compared with it
are the nutmeg plantations of Banda and the cocoa plantations of
Trinidad in the West Indies. I only wish that you could stroll with me
some morning early here, when the sun is up, but not yet high or hot,
when the dew is sparkling on every leaf and blade of grass, when the
birds are singling flutily and the flocks of gray and crimson lories go
swishing out from some tree, when the air is redolent with the perfume
of spice and flowers, when the coloring of the foliage close around, of
the fruit and bamboo forests on the volcanoes slope, of the sky overhead,
of the distant cones of Tidore and Makian are brightest in the rays of the
rising sun, then surely you would think yourself in a second Garden of
Eden.

However, earnest Henry Moseley's scientific interests were not
so easily derailed by earthly paradise. He was determined to climb
the peak and did so in the company of four Malayan guides and
one of *Challenger's* sub-lieutenants, A. F. Balfour. They climbed
through the cultivated lands of the lower slopes after spending the
night in the house of a Dutch plantation official. From two to four
thousand feet they ascended through tracts of verdant deep-green
forests before entering a region of dense reed. At 4,800 feet they
reached an ancient outer crater of Ternate, the quiet home to nothing but bushes, tree ferns, deer, and wild pig. Above them, though,
were fields of recently erupted lava. Bypassing these, they eventually
came upon the active caldera. As Moseley and Balfour began their
final ascent, the native guides' nerve failed them and they turned
tail and ran. Moseley wrote, "We were told afterwards that they
have strong superstitious fears concerning the volcano, and believe
that if anyone climbs the terminal cone, a terrible eruption and
earthquake are certain to ensue." Moseley appears to have harbored
certain doubts of his own about the wisdom of entering the caldera,
because he wrote wryly, "It appears as if there might be some real
risk in the ascent."

The view inside the cone was of another world. Billowing
clouds of steam and smoke drifted and eddied, obscuring then
revealing the tormented reddish rock of the opposite wall. Moseley

and Balfour tried to climb down toward the caldera's floor but the noxious brew of gases billowing around them forced them back. "It was only possible to descend about twenty yards into the crater," Moseley wrote later, "and even then the vapors inhaled were very trying. Steam and acid vapors issued from cracks everywhere, decomposing the lava amongst which we passed. In most of the cracks were small quantities of sulfur." It was some compensation, though, that the view from the top of Ternate was truly spectacular, and Moseley and his companion enjoyed a sight not often seen, because the volcano was seldom climbed. Below them and to the south, the island volcanoes of the Dutch East Indies stretched back along their course, an arrow pointing toward Australia and a crown jewel of their own empire.

But it was time to leave this ailing remnant of another nation's imperial ambitions, and on October 17, 1874, they left the Moluccas and headed for the Philippines. They arrived in Zamboanga late in the evening of the 23rd. Campbell enjoyed the town, being impressed with the diversity of the Chinese shops there as well as the quality of the wares. "Everything can be got and of the best quality," he wrote, "from portmanteaus to pate de foie gras, from bewitching velvet—silver, gold and silk embroidered—slippers to Bass's bottled pale ale. . . ." The romantic and impressionable Herbert Swire was particularly taken with the diminutive ladies of the town. "These young ladies," he wrote, ". . . had such slender waists, such graceful limbs, in fact were so perfectly formed, that one would never think of saying 'she is too short for a beauty.'"

At daybreak on Monday October 26, 1874, they left Zamboanga and steamed out into the Basilan Channel bound for Manila, the capital of the Philippines. After pausing briefly at the island of Ilo-Ilo, they reached Manila on November 4. For the first time in several weeks, Campbell was disappointed with a port of call.

> Manila, being the capital of the Spanish Philippines and a great cigar making place, I expected to find some resemblance there to its great rival Havana, but there is absolutely none. No large and crowded cafes with bands playing in the galleries above and luscious drinks below; no

volantes full of dark eyed Creoles and Spanish ladies driving around the plaza, with fireflies sparkling in their hair; nothing in one word of what makes Havana a charming place. . . . I shall not describe to you this fourth rate Spanish town, where the only amusements are to eat prawn curries in a hotel, drink in a poor café, and drive in open carriages. . . . The mere fact of saying that these islands are Spanish is as much as saying that they are a misgoverned, backward state. No enterprise and only three "open ports" throughout the length and breadth of the great archipelago. Earthquakes are constant, almost unceasing. In the old city of Manila are the picturesque ruins of a magnificent old cathedral, shattered by a tremendous shock some years ago.

Much to his relief, they left Manila on the 11th of November, arriving in Hong Kong on the 16th, where a tremendous shock awaited the ship's company: Captain Nares had been ordered back to England to take command of the forthcoming Arctic Expedition. Accompanying him from *Challenger* would be Lieutenant Pelham Aldrich. It was a bitter blow for Scientifics, officers, and crew, but especially for Wyville Thomson. "Professor Thomson was in a great way about it," wrote Joe Matkin, "and talked about throwing up the whole affair and coming home, but the captain persuaded him not [to]. . . ." In Nares's honor they threw a grand farewell dinner during which he made a speech saying how sorry he was to go and that he owed his promotion to the zeal of his officers and men. It was with a heavy heart that they said goodbye to the man who had brought them from England to the Far East with such facility and dispatch. But, at least one man aboard harbored lingering doubts about the wisdom of sending Nares to the Arctic. "I don't think Captain Nares is quite strong enough for such a voyage," wrote Matkin, "he suffered from "Rheumatics" on the Antarctic trip and he is rather a timid man I think—not enterprise enough for such a command."

Captain Frank Thurle Thomson of HMS *Modesty*, and Lieutenant Carpenter of HMS *Iron Duke*, both frigates on the China station, were to be their replacements. Thomson was due to arrive within days from Shanghai and his arrival was viewed by some aboard with a trepidation succinctly summed up by Matkin, who wrote, "He bears a bad name for tyranny on this station."

They stayed in Hong Kong for more than two months and on January 6, 1875, headed south again toward the Philippines. South of Mindanao they turned east for Humboldt Bay on the coast of New Guinea. Finding the natives hostile they elected not to prolong their visit. Instead they pushed on for the Admiralty Islands where, arriving on March 3rd, they found a friendlier reception as well as some of the most exquisite native artwork that they had yet encountered. The woodcarving was especially good and because the natives were keen to barter, members of the ship's company acquired several examples. They stayed in the Admiralty Islands, anchored in the newly named Nares Harbor, for a week, after which they were pleased to be on their way. The islanders, although friendly, lived a life pinned between impenetrable rainforest in the center of the island and the vast Pacific Ocean on the other. Their life was narrowly circumscribed and completely unattractive to denizens of an empire on which the sun never set.

William Spry summarized the feelings of many of the crew, "In the natives of Humboldt Bay and Nares Harbor, we had had an opportunity of seeing man uncontaminated by civilization, and free to follow the bent of his own free will utterly untrammeled by society or customs. Under such conditions man is a degraded animal, and the noble savage as great a myth as the elixir of life."

For all that, the Scientifics found their time in the Admiralty Islands especially rewarding. Moseley, with his appetite for anthropology, found it among the most interesting places they had yet visited. He couldn't have guessed that within a century the Admiralty Islands would be converted into American military bases for the U.S. war effort in the Pacific and the native culture totally swallowed up.

The next leg of the voyage infected all aboard with its tedium. Moseley wrote, "A fact often brought home to me before, during the *Challenger's* cruise, was tediously forced on our notice on this voyage to Japan, namely that the inmates of a sailing ship on a long voyage suffer far more from too little than from too much wind. We were constantly becalmed, and our steam power being only auxiliary, and coal being short, we had to lie still and wait, or creep along occasionally only at the rate of a mile an hour."

Swire noted the tedium, too, and also wrote of its consequence: that tempers were frayed and faces sullen in the wardroom. However, as they approached the Mariana Islands of the western Pacific, he was just about to encounter the one incident on his long journey around the world that would in future years afford him the greatest pleasure. On March 23, 1875, 13 days after leaving Nares Harbor, soundings indicated a depth of 4,475 fathoms or about 27,000 feet. This staggering abyss, now known to be almost 7 miles deep, was by far the deepest part of the seafloor that *Challenger* encountered. To honor both the occasion and the popular young sub-lieutenant, the Scientifics named it Swire Deep (although sadly, after they had returned to Britain, the name was later changed to Challenger Deep).

We have already mentioned that subduction zones are not restricted to areas where an oceanic tectonic plate is sliding under a continent. Oceanic crust can also be subducted under other oceanic crust, and the Marianas are just such a region. Here the Pacific Plate is being subducted under the Philippine Plate and, like the islands of the Dutch East Indies, a chain of volcanoes has formed above the subducting slab. There are 14 islands in the Marianas chain, which stretches for a thousand miles from Guam in the south to Farallon de Pajaros in the north. The Marianas, like all volcanic chains above subducting slabs, is arced in a vague crescent shape. Why is this? To answer the question, simply imagine that planet Earth is a ping-pong ball. In this mental experiment, press the ball with your thumb to represent the subducting slab. The result is a circular dimple. On a larger scale, volcanic arcs form around the margins of similar dimples on the surface of the Earth. These are the regions where subducting slabs slide back into the interior of the planet.

The *Challenger* expedition had discovered the deepest part of the world's ocean at the bottom of one of these arcs. The two thermometers they sent to the bottom of Swire Deep were retrieved broken by the stupendous pressure of 7 miles of water. Yet when they recovered the sounding tin, it contained traces of mercury from one of the broken thermometers. The implication was obvious: So placid were the bottom-water currents that wherever in the water column the thermometers broke, their mercury dropped so perfectly vertically that the sounding tin landed in the same spot. Many of the crew wondered what it could be like in that unimaginable world, so close—only 7 miles away, after all—yet so resolutely remote. But perhaps the most extraordinary thing about the Swire Deep is that within 90 years, for the first time in its long history, its endless night would be dispelled by a personal visit from the sunlit realms above.

BUBBLES IN THE DEEP

At 4:56 in the afternoon of January 23, 1960, two American navy jets screamed over the water only a handful of miles from the site of *Challenger's* position 85 years before. But bobbing on the swell below them as they banked away was not the wood and iron bulk of *Challenger* but an alien-looking bubble of aluminum, below which was suspended a massive iron sphere. It was the *Trieste*, the first submersible to reach this deepest part of the ocean floor. Aboard were two men, American naval lieutenant Don Walsh and Swiss scientist Jacques Piccard. Piccard was the son of the man who had invented this strange craft, a bathyscaphe. They had just completed an extraordinary journey, 7 miles straight down to the bottom of the Swire Deep (Challenger Deep, as it was by then called) using the revolutionary new technology of this greatest of deep-sea submersibles. But in truth it was a journey that had started just a year after *Challenger* returned to Spithead in 1876.

William Beebe was born in 1877 in Brooklyn, New York, and from his very earliest days had an overwhelming interest in natural history. His family often visited the American Museum of Natural

History in New York. So frequent were Beebe's visits, and so obvious and sincere his enthusiasm, that he soon struck up a friendship with the noted paleontologist Henry Fairfield Osborn. Osborn was not only the president of the New York Zoological Society (and, therefore, the museum's chief executive officer) he was also a full professor at Columbia University and, at the time he befriended Beebe, one of the most influential men in the natural sciences in America. He was also one of the prime movers in the plan to build a zoological park for the City of New York.

In 1896, after Beebe graduated from East Orange High School, Osborn arranged for him to be accepted into Columbia as a special student. Although he spent the next three years there, despite an agreement with Osborn that he would receive school credit for his work for the museum, Will, as he liked to be known, never officially graduated. For the rest of his life Will himself never disabused anyone of the notion that he had received a bachelor's degree in science from Columbia, but it was not until the late 1920s that he received honorary doctorates from Columbia and Tufts Universities. It was also at about this time that he began to shift his interests away from ornithology into marine natural history. With the aid of a crude helmet, he dived in the ocean and trawled from his yacht, *Arcturus*, just off the coast of Bermuda.

As a now wealthy scion of American intellectual society Beebe had friends in high places, one of whom was Theodore Roosevelt. Roosevelt listened sympathetically one evening at dinner as Beebe expressed his frustration that so many fascinating specimens were arriving on deck destroyed by decompression (a problem that had also dogged *Challenger's* Scientifics). Decompression occurs when animals from the abyss are brought to the surface. Dissolved gases in the animals' bodies, immensely pressurized in the deep, expand and destroy them as they are brought up. Roosevelt sketched on a napkin a vessel that could go to the bottom of the sea. It was vaguely spherical and looked like a heavily armored bathtub that was sealed at the top. Even in that early version Roosevelt's vessel had many of the features that would later come to be associated with the craft known as the bathysphere.

Beebe decided to pursue the idea of building a bathysphere and in 1926 the *New York Times* published an article publicizing his interest. As a result, his office was inundated with designs for the deep-sea vessel that ran the gamut from a Jules Verne fantasy to plausible design. It is here that the strange and enigmatic character of Otis Barton enters the story.

Barton, an American engineer, was an intensely private man and even today there is virtually no information about his personal life. From an early age he was interested in investigating the deep sea. As early as 1917 he made his own primitive wooden diving helmet complete with glass windows and, weighing himself down with rocks and breathing air pumped down to him from a bicycle pump operated by a friend, used it to investigate the rocky bottom of Cotuit Bay near his home in Massachusetts. Barton's interests, unlike Beebe's, were more in overcoming the engineering prob-lems than in natural history. However, recognizing that deep-ocean exploration would inevitably be driven by biological concerns, he studied both engineering and natural sciences at Columbia University. Unlike Beebe again, Barton did not need a conventional job, because he had inherited a large amount of money on the death of his grandfather.

Like much of the American public, Barton was a great fan of Beebe's exploits and read his books avidly. When he heard of Beebe's plans to explore the deep ocean, Barton wrote to him expressing interest and informing him that he had a design for a submersible. Beebe ignored the letter because by now he was mightily tired of crank mail containing unworkable designs. Then by a strange stroke of good fortune for science, a mutual friend who knew of their shared interests persuaded Barton and Beebe to meet. They got together on December 28, 1928, in Beebe's New York office and immediately hit it off. Beebe was impressed by Barton's design and the great amount of thought that had obviously gone into it.

Barton's vessel was spherical, an arrangement that combined the maximum strength (because of the even distribution of pressure) with the minimum weight.

They agreed to work together, with Barton funding the design and manufacture of the sphere himself. The New York Zoological Society and National Geographic Society provided additional funding, the latter gaining exclusive publication rights to their expeditions for the next 10 years. The bathysphere, as it came to be known, was manufactured by the Watson-Stillman hydraulic machinery company of Roselle, New Jersey.

It was 5 feet across, had cast iron walls one-and-a-half inches thick, which Barton calculated would withstand the pressure at four-fifths of a mile, and could hold two people. The sphere was tethered to the mother ship, *Ready,* by a single non-twisting cable seven-eighths of an inch thick and 3,500 feet long. It had a breaking strain of 29 tons. Electricity and communications ran through a metal-cored rubber cable beside the steel support cable and entered the sphere through a specially constructed port.

Entry and exit for the two occupants was through a narrow circular opening whose hatch was secured by 10 large steel bolts with another wing nut placed centrally to allow for quick opening in the case of an emergency. Opposite the door were three circular windows constructed of fused quartz. Each was 8 inches in diameter and 3 inches thick and they were manufactured by the General Electric Company to a design by Dr. E. E. Free of New York University. Free, an expert in the optical properties of materials, chose quartz because it allowed light waves to pass through the glass without distortion. Although five quartz pieces, each costing $500, were initially constructed, only two passed Barton's stringent fitting and pressure tests, so a steel plug replaced one of the windows on their first dive.

The interior of the bathysphere was cramped and spartan. Two small oxygen tanks, placed on either side of the windows, kept the air fresh for up to 8 hours. Above the windows were wire-mesh trays filled with soda lime and calcium chloride to remove exhaled carbon dioxide and moisture from the air. The entire apparatus had cost Barton a colossal $12,000. In early May 1930 all was finally ready and Barton sailed with his 11 tons of equipment for Nonsuch Island, Bermuda where Beebe based his operations.

The early dives of the bathysphere were unmanned but not without incident. On the first dive, on June 3, 1930, the massive steel supporting cable and rubber utility cable became so badly entangled that the cable could not be rewound on its reel. They feared that they could not retrieve it but slowly the winch pulled up the massive bathysphere while, inch by inch, the crew pushed the rubber hose down the steel hawser like a vacuum cleaner cord that had become entangled around a garden hose.

When the sphere arrived back on deck they found that the "nontwisting" cable had rotated about itself no fewer than 45 times. On another unmanned test the sphere took on water at depth and when the central bolt was loosened, a stream of water shot out with enough intensity to cut in half anyone who ventured into it. Almost immediately after that, a side bolt came loose and blasted across the deck with the velocity of a cannon shell, gouging a two-inch curl of cast iron from one of the winch housings. It was a sobering reminder of the colossal forces that lurked in the deep.

Eventually all these technical difficulties were overcome and Beebe and Barton, in what must remain as one of the most intrepid explorations of the twentieth century—certainly ranking with Armstrong and Aldrin's later foray on an equally alien world—were ready to take on the silent landscape. At 1 P.M. on June 6, 1930, *Ready's* captain signaled to the bathysphere's crew chief that he was ready to begin the dive. As the capsule sank below the surface Beebe wrote, "We were lowered gently but hit the surface with a splash that would have crushed a rowboat like an eggshell, yet within we hardly noticed the impact until a froth of foam and bubbles surged up over the glass." At 400 feet—the greatest depth achieved by the submarines of that era—Barton exclaimed and Beebe turned to see water trickling in from the seal around the main hatch; but rather than cancel the descent, Beebe requested that their rate of dive be increased. At 525 feet they passed the greatest depth attained by a deep-sea diver in an armored depth suit. They were now deeper than anyone had ever gone before, truly in uncharted waters. At 600 feet Beebe called a halt so that he and Barton could assess the problem of the leaking hatch. Although it seemed no worse, some sixth sense warned Beebe not to continue farther that day. He later wrote, "Some mental warning which I have had at least a dozen critical times in my life spelled bottom for this trip."

Four days later, after Barton had packed soft lead into the groove around the leaky door's periphery, the expedition was ready for another attempt. The sphere was lowered empty to 2,000 feet, and brought back to the surface, returning unsnarled and with no evident leaks. On June 11 the team decided that they were ready to go again but this time determined to reach serious depths. At 10:00 A.M. they were winched overboard.

Beebe, by this time an accomplished writer, and mindful of his funding by the National Geographic Society, wrote eloquently and accurately of their descent. They passed through regions where long strings of salp drifted slowly past, "lovely as the finest lace, while schools of jellyfish throbbed their energetic way through life." At 800 feet Beebe got his first glimpse of a hatchet fish. At 1,000 feet the deep ocean had begun to resemble the night sky, as schools of glowing creatures drifted past the descending capsule. Like the crew of *Challenger*, Beebe noticed that this bioluminescence was not static but rather shifted and changed. Inside the sphere they found that that were now sitting in a pool of water as the balmy Bermudan air condensed on the cold walls of the sphere.

At 1,250 feet the sphere entered an apparent dead zone where no life was to be seen. For a few startled moments it seemed to the explorers like a strange echo of Forbes's azoic theory that Charles Wyville Thomson and William Carpenter had confounded 60 years before. At 1,400 feet they stopped. Both were now uneasy. Beebe wrote, "The water below looks like the black pit-mouth of hell itself." The pressure at this depth was a staggering 650 pounds per square inch and the view ports were now experiencing a pressure of 9 tons but still the bottom of the ocean was thousands of feet below them. Beebe and Barton had approached closer to the silent vastness of the seafloor than any man alive. They returned from a recorded depth of 1,426 feet to a rapturous welcome.

On September 22, 1932, Beebe and Barton set a new depth record of 2,200 feet. Beebe described "a feeling of utter loneliness and isolation akin to those which might grip the first to venture upon the moon or Venus" as he and Barton dangled alone and helpless in the trackless depths of the Atlantic Ocean. On this series of dives, Beebe and Barton made broadcasting history by transmitting a narrative of their adventures to a spellbound world via NBC radio.

In 1934 Beebe carried out his final series of dives, which achieved a depth of 3,028 feet, more than half a mile down. That was close to the sphere's limits. It had served them well but at this depth the cable was almost completely unwound on its reel and the captain was terrified that it would slip off the drum and drop the explorers to the bottom of the ocean. Doubtless the sphere would have failed before reaching the seafloor, crushing his friends flat in a second as the pressure of the deep snuffed out the upstart bubble that had invaded the silent landscape, but the thought of Beebe and Barton dying a lonely, long-drawn-out death of asphyxiation must have haunted him too. He allowed them to stay at this depth for only a few minutes before hauling them back. There was a tense moment when a sharp snapping sound rang through *Ready* but it was only one of the guide ropes that guided the steel hawser back onto the drum giving way.

Despite the radio broadcast, the dives were much more than mere showmanship. After each series of dives Beebe compared his direct observations with those he had made during trawling and dredging operations. He was able to show that certain regions of

the deep were much more populated, particularly by large vertebrates, than tows at different levels would suggest. Clearly the fauna
of the deep could see dredging nets coming, and the tows made by
Challenger could not have brought up many organisms that were
fully representative of the life of the deep.

In 1934 Beebe summarized his dives and observations in a
book, called *Half Mile Down*, which remains a classic. Why he halted
his dive program in that year remains something of a mystery.
Perhaps, at 57, he felt too old to continue to work in the cramped
confines of the sphere. His retirement from the field marked a hiatus
in the manned exploration of the deep ocean, which was not
resumed until after the Second World War. The craft that resulted
from this later phase was the one that eventually lighted the eternal
darkness at the bottom of Challenger Deep

In the immediate aftermath of the Second World War, Otis Barton
decided to build another sphere and resume exploration. He called
the newly constructed sphere a benthoscope and started testing it
in August 1949 off the coast of California. Although he set a new
record with a dive to 4,500 feet, that signaled the end of tethered
deep-sea diving vehicles. To go deeper, progressively thicker walls
were needed, and longer cables. For each foot of depth gained, the
craft became heavier until eventually no surface vessel could support
its weight. Something more versatile was needed, and even as Barton
was testing the benthoscope, a Swiss scientist named Auguste
Piccard was putting the finishing touches to a deep-sea diving vessel
with no need of tethers, the bathyscaphe.

The biggest difference between a bathysphere and a bathyscaphe is that the bathysphere is heavier than water and the
bathyscaphe is lighter. While the bathysphere dangles underwater
passively at the end of the metal rope, the bathyscaphe, whose living

chamber is still a heavy sphere like that of the bathysphere, is attached to a massive thin-skinned steel tank filled with gasoline. Because gasoline is lighter than water the craft has an innate buoyancy. Only a thin-skinned chamber is needed to contain the gasoline because, like all fluids, it is effectively incompressible.

Auguste Piccard conceived the bathyscaphe in the 1930s but became distracted by the allure of high-altitude ballooning. But in the late 1930s, after meeting Beebe at the World's Trade Fair in Chicago, Piccard decided to build his craft. He was encouraged by the King of Belgium, who helped secure funding from the Fonds National de la Recherche Scientifique (FNRS), the Belgian scientific research funding council. The bathyscaphe construction program was put on hold during the Second World War but was taken up again immediately afterwards.

The "scaphe" (as it is commonly abbreviated) FNRS 2, was tested on November 3, 1948, off Dakar, North Africa. It suffered damage in the heavy Atlantic swell and 6,000 gallons of gasoline had to be jettisoned, triggering a funding crisis. Auguste Piccard turned to the French government for help. Although interested, they were unwilling to countenance a foreign scientist as the head of the research program. Devastated but desperate to prove the worth of his idea, Piccard handed his brainchild over to the French.

At about this time his son Jacques was finishing university in Trieste, Italy. A professor there heard of Auguste's plans, became interested, and offered to help him find funding for a new vessel provided that it bore the name *Trieste*. Overwhelmed, the Piccards agreed and by 1953 the *Trieste*'s pressure sphere was beginning to take shape in the Terni steel mills just north of Rome. The new pressure sphere was made of a forged steel alloy and was stronger than the cast steel used in FNRS 2. With a diameter of 7.25 feet, and walls 3.5 inches thick, the new sphere could resist a pressure of 10 tons per square inch. The windows of the new vessel were made, not of fused quartz, but of a radical new flexible plastic called plexiglass. The flotation tank contained 22,000 gallons of gasoline, imparting enormous buoyancy to the craft. Electromagnets at either end of the craft held 9 tons of shot to be released before ascending. The sheer size of the craft made it extremely hard to maneuver and its maximum speed was only 1 knot. Unlike Beebe, who kept the inside of his craft as spartan as he could, Piccard filled his crew compartment with equipment.

So now there were two bathyscaphes: the original that the French had wrested from Piccard, the other built by an unlikely Italian-Swiss consortium. The two were engaged in a race to the bottom. Barton's benthoscope record was broken in the summer of 1953 when, on August 1, *Trieste* reached the bottom of the Mediterranean, a depth of 10,390 feet or almost 2 miles. To continue their race, both parties would now have to move to deeper ocean basins.

In January 1954, just off the Cape Verde Islands, the French, in FNRS 2, reached a depth of 13, 287 feet, the floor of the Atlantic in that region, and had just time to survey a field of swaying sea anemones before an electrical fault caused the electromagnets holding the ballast to trip and send them back toward the surface. With true gallic exuberance, the French crewmen aboard, Willm and Houot, celebrated their achievement by opening a bottle of wine and feasting on a packed lunch as their released machine shot them back to the surface.

In the mid-1950s Jacques Piccard was in the U.S. trying to drum up support for a more extensive and scientifically oriented program for *Trieste*. He was gratified and a little surprised by the level of support he found. But this was the era of the Cold War. Sputnik would soon be launched and the American people were just beginning to wake up to the new and terrifying realization that their supposed technological supremacy was being seriously challenged. America planned to conquer "inner" space with the same fervor with which they were attacking outer space, even if the competitor was not the Soviets but the French.

In February 1957 the U.S. Office of Naval Research (ONR) drew up a contract asking the Piccards to carry out a series of test

dives in the Tyrrenhian Sea between Sardinia and Italy. From July to
October of 1957 *Trieste* made 26 test dives carrying a range of
underwater specialists. These tests proved to the Americans the value
of bathyscaphes and ONR decided to buy *Trieste* outright. The
Piccards preferred some form of leasing arrangement but ONR
insisted on the purchase and eventually they capitulated. *Trieste* was
relocated to the Naval Electronics Research Laboratory in San
Diego, California, close to the famous Scripps Institution of Oceanography at La Jolla. Infected with space-race fever, the Americans
were bullish about their new technological acquisition. Their goal
was the conquest of the deepest spot on Earth, the Challenger Deep
of the Mariana Trench, off the Pacific Island of Guam, 8,000 miles
to the west, discovered 85 years before and originally named for a
young sub-lieutenant on an equally epic voyage.

Quietly, so as not to arouse the suspicions of the press, Jacques
Piccard began constructing a new pressure sphere. He approached
the Krupp Steel Works of Essen, Germany, which had been the
principal armaments manufacturer of Nazi Germany, with the
project. Although Krupp had lost most of its heaviest steel-making
equipment to Yugoslavia as war reparations, the engineers there were
confident that they could build the new sphere. It would be made
in three parts and joined together so precisely that, they claimed, it
would be as though they had forged the sphere in one part. To
withstand the staggering pressure of 9 tons per square inch at the
bottom of Challenger Deep, the new sphere's walls were 5 inches
thick. It weighed 13 tons in air, 8 in water; the *Trieste's* float had to
be modified to hold more than 40,000 gallons of gasoline.

Trieste departed San Diego on October 5, 1959. Between
November 1959 and early January 1960 she underwent test dives in
the waters off Guam, reaching a depth of 23,000 feet—slightly more
than 4 miles. The "big dive" was set for January 23, 1960, and
Lieutenant Don Walsh and Jacques Piccard were selected as the
crew. On January 22, *Trieste*, towed by the USS *Wandank*, reached
station above the deepest part of the Challenger Deep. Echo-

sounding revealed the depth there, at 11° 19′ 7″ N, 142° 12′ E, to be 35,700 feet, about 10,000 feet deeper than the part of the Deep the *Challenger* had sounded.

Walsh and Piccard began their 14-hour roundtrip at 8:23 A.M. on the morning of the 23rd. They ran into problems at only 340 feet when they encountered the thermocline layer where temperature and water density change rapidly. The balance between the *Trieste's* positive buoyancy and the ballast needed to take it gently down to the seafloor was so delicate that the craft simply bounced off this layer. Piccard had to jettison the gasoline in order to penetrate the layer but in doing so was permanently throwing away a portion of the positive buoyancy needed to return them to the surface. Three more times *Trieste* had to fight its way through other thermal layers and each time Piccard had to jettison more precious gasoline. With two days worth of oxygen on board they could simply have waited for the gasoline to cool, which would have increased its density and then their journey would have resumed automatically. But the aching cold of the deep ocean was already beginning to penetrate the sphere and Piccard and Walsh were anxious to return to the surface before nightfall.

Navigation was crucial, too. The slot in the ocean floor they were aiming for was less than a mile wide and with the craft's immense float, they had virtually no maneuvering ability. By 11:30 in the morning the divers had reached a depth of 27,000 feet— more than 99 percent of the water in the world's oceans was now above them. At 32,400 feet, a massive explosion rocked the bathyscaphe, but they continued their descent anyway, aware that something outside the pressure sphere had broken. But by now they knew that the structural integrity of the mighty Krupp sphere itself was unlikely to be compromised. If it were, they would have known nothing about it. At a pressure of more than 6 tons per square inch, death, in the event of a structural failure, would have been instantaneous.

At 12:56 P.M. a signal registered on the paper scrolling through the echo sounder. There were only 42 fathoms, 252 feet, to go before they hit bottom. Piccard began releasing steel shot to slow their descent and then, with only 3 fathoms to go, a light-colored ooze came into view. The depth gauge showed 37,800 feet, a depth that exceeded their expectations. They were later to learn that the gauge had been calibrated in fresh water—but their true depth was still a staggering 35,800 feet.

In this amazing place the proof of Wyville Thomson's and Carpenter's refutation of Forbes's azoic theory was immediately apparent, because the first thing Piccard saw as he stared out of the window was a flat fish, its two large eyes staring at him. As the bathyscaphe gently touched down, the fish rose languidly and swam away. The two divers shook hands; they had made it. Then, as Don Walsh stared through an up-angled view port the cause of the explosion became clear. A plastic window in the connecting passage between the sphere and the exit had cracked under the colossal pressure.

After only 20 minutes at the bottom Piccard dropped ballast to start their slow ascent. Gradually they made their way back toward the surface, eventually reaching a maximum vertical velocity of 5 feet per second—almost twice as fast as a commercial elevator. Their rate of ascent was temporarily slowed as *Trieste* broke through the thermocline layers that had caused so much difficulty on the way down. Even now, the return journey was not without worry, because after many hours of abyssal temperatures the gasoline in *Trieste's* float was the same temperature as the water of the Challenger Deep and the craft's innate buoyancy, compounded by the venting needed to get them to the bottom, was at a minimum. But they reached the surface safely at 4:56 in the afternoon of January 23, 1960. It was then, as they clambered up the scaphe's ladder to the fresh air and bright yellow sunlight of the western Pacific, that the two navy jets screamed overhead, dipping their wings in synchronized salute to the first humans in history to reach, in person, the deepest part of the silent landscape.

Challenger Deep—Swire's Deep as it had been originally named—had finally been seen by human eyes.

THE HOUSE OF THE RISING SUN

The good wind that had sprung up 10 days out of Nares Harbor failed again and *Challenger* inched closer to Japan under all plain sail, to conserve their dwindling fuel reserves. Once again tempers frayed but this time there was nothing to be done, no record "deep" to be found to break the monotony and remind them of the importance of their voyage. It was not until they were almost in sight of Japan on April 11, that the wind picked up and they then made good time up the Gulf of Yedo toward Yokohama, accompanied by sharks, porpoises, and dolphins that had come out to greet the ship. They landed on the 12th and all aboard were immediately impressed with this new land.

Japan had been opened to foreigners only recently, when the empire was reinstated after two centuries of rule by the feudal war-lords known as Shoguns. The last Shogun had been persuaded to resign as recently as 1867 and only since then were certain cities, the treaty ports, open to Europeans. But Japan was embracing contact with new cultures, as well as the numerous trade possibilities, with such enthusiasm that permits for more widespread exploration were not hard to find if one had a good reason and an accredited guide. "You know that Japan is not yet 'open' to foreigners," wrote Campbell, "only in the 'treaty ports' and defined boundaries around them and in 'concessions' in the towns of Yedo and Osaka, can foreigners come, go, or stay as they list. We, and everybody traveling outside these limits have to get passports which are portentous look-ing documents and procured through the legations." For a vessel whose goals were as quintessentially scientific as those of *Challenger*, there was no question that the officers and Scientifics would be given free rein. Thus began two of the most pleasant months of the entire cruise.

Henry Moseley soon made friends with a Mr. Dickens, a Yokahama-based barrister with a keen interest in natural history. They struck up a firm friendship and together visited Kobe and Kyoto. Moseley's eye for anthropology remained as keen as his eye for terrestrial biology. He observed how important religion was to the Japanese and how prevalent still was the practice of pilgrimage. "Pilgrimages are extremely popular in Japan" he wrote. "On the journey along the Tokaido (the east sea road), the road was thronged with pilgrims going to the ancient shrine of Ise, the oldest temple in Japan of the Shinto religion, the ancient state religion of the country, of which the Mikado, descended from the gods is the supreme head." He was delighted to discover the Japanese penchant for the written word and spent a happy few days in the booksellers' quarter of Osaka. But it was the landscape that attracted his most lavish praise as they traveled. "The land along the road is in the very highest culture. A great deal of it was covered with yellow-blossomed crops of rape, whilst here and there were wheat crops. The straightness of the lines of planting, and the regularity of their distances from one another, was such as I have never seen approached elsewhere in any form of agriculture."

Herbert Swire's enthusiasms were more mundane and after a month at sea his libido was, as ever, stirring. "The girls, or rather I should call them Musume (pronounced Moo-ze-me) because I think it is a prettier word, are the neatest little ladies I have ever seen and very many of them are exceedingly pretty. They powder their necks and foreheads and rouge their lips and make no bones about it either." In Britain in those days make-up was still a rarity so it is easy to see how Swire might have been stirred by such palpably erotic behavior. The Geisha ethos would have been as far away from that of polite London society as the dark side of the moon. Swire was clearly torn between loyalty to the homegrown female and these dusky beauties of the Orient, because he added hastily and with breathtaking condescension, "Of course, I am not going to compare English ladies with these for, setting aside the difference in

brainpower etc there is an immense difference in personal appearance greatly in favor of the English beauty to an English eye." He
was every inch the young empire builder away from home and, like
military men through the ages, not above taking his chances, "But
the Musumes are so neat and clean, so healthy looking and so good
tempered and above all so weak and helpless on account of their
delicate build that one really gets quite spoony on them. I don't get
spoony on any one more than another, but I must say I'm very
spoony on them all."

But they were also still in the land of earthquakes. Swire wrote,

> [Because of] the great frequency of earthquakes in Japan they have
> adopted a unique system of building their large houses and temples (all
> being of wood). The roof is first made, and everyone knows what a solid
> looking affair the Japanese roof is; the roof is then raised by some means
> from the ground and rested upon wooden pillars which fit into it; the
> lower end of the pillars, instead of being firmly fixed into the ground,
> are merely placed on smooth stones, raised a little above the ground, so
> that when an violent earthquake occurs the temple oscillates freely. . . .

Campbell was so keen on Japan that he devoted an entire
chapter (entitled "A Peep into Japan") to it in his *Log Letters*. However, the word "peep" acknowledges his realization that despite the
two months that they stayed, they only scratched the surface of this
complex place and culture.

Below decks Joe Matkin reveled in the complexity and richness of Japan even as he wrestled with his grief. Awaiting him when
they had arrived was the news that his beloved father had passed
away some four months before, while *Challenger* was still in Hong
Kong. He wrote to his mother, saying,

> I am thankful to think that Father lived long enough to see us all pro-
> vided for and properly educated. Few children in our station of life have
> had so much spent on their education and start in life. . . . By the time
> you get this we shall be 'Homeward bound' and you will be able to
> count the months instead of years as they fly past. I have fully deter-
> mined to leave the Navy when I get back; there will be nothing to
> hinder me; I shall have a little money and what is better a good character!
> and I hope we shall all be settled down in England and have many happy
> years yet in the old home.

Toward the end of their time in Japan *Challenger* visited the inland sea, spending time in Osaka before returning to Kobe. Shortly thereafter they returned to Yokahama and then left Japan on June 16, 1875 for the Sandwich Islands or, as they are better known today, the Hawaiian Archipelago. As they left Japan, Campbell summarized the feelings of all on board when he wrote, "Let me advise all those who wish to travel and find real novelty of scene, combined with comfort and cleanliness, to visit Japan."

Dreams of Big Science

Yokahama, Japan, June 16, 1875, 35° 28′ N, 139° 38′ E, to
Portsmouth, Great Britain, 50° 48′ N, 1° 05′ W

From Japan, *Challenger* headed east into the immensity of the North
Pacific. The dredging hauls were poor, because the seafloor in that
region was rocky and barren. For those on board this leg was as
dreary as their voyage from the Admiralty Islands to Japan had been.
Many had had their fill of the cruise and none felt this way more
keenly than Herbert Swire: "For nearly one mortal month we have
been at sea without one sight of land and only once chancing across
a ship. I am sick of it. Very much so. . . ."

At 3:00 P.M. on July 27, 1875, *Challenger* anchored outside the
reef at Honolulu. Reactions to the town were mixed, Joe Matkin
finding it "civilized" but Herbert Swire writing, "I was much disappointed on sighting these islands. From reading various books
having reference to the group, I fully expected to see islands
luxuriant with verdure or at least the beauty of an ordinary tropical
coral island. Instead of this, however, I found Oahu, the island upon
which stands the capital Honolulu, to be an elevated ridge of barren
rock, stretching about 30 miles in a NW and SE direction."
Campbell's account concurred and they were all glad to move on a
few days later for the "big island" some 30 miles to the southeast.

They found Hilo Bay much more rewarding than Honolulu,
although still not what they, with their experience of Indonesia,
would call a tropical paradise. All the men eagerly anticipated the
forthcoming visit to Mauna Loa to see the active volcano although

initial impressions, gained from the harbor, were not especially exciting. "Owing to their gradual ascent," wrote Herbert Swire, "these two mountains look quite insignificant. . . . Mauna Loa especially looks ridiculously small, only a molehill, as many of us remarked."

But this disparaging impression of Mauna Loa changed when they started to climb it. Campbell and Moseley set out at two o'clock in the afternoon of the day they arrived in Hilo but did not arrive at the hotel on the rim of Kilauea until 1:30 in the morning. But the view during the final phases of their ascent was worth it. "Presently a red glow appears among the clouds on our right, increases and then the clouds which before had covered it melt away and Mauna Loa reveals its long, low lying summit from the center of which a great column of lurid light and smoke is flaring. . . ."

The next morning the rest of the party joined them. The hotel was of a surprisingly high quality despite its location right on the edge of an active volcano. The plumbing arrangements, however, perplexed the travelers. A grass hut had been erected some distance from the hotel over a crack in the lava through which steam issued. The steam condensed in the grass and, together with any rain that might have fallen into the grass roof, ran back into the hut, where it was collected and stored in tanks. Over the tanks was a hand pump to pump the warm whiskey-colored water into a bathing tub.

Farther down the hill was an even more dangerous bathing contraption, a sulfur-vapor bath. A box had been erected over a volcanic vent from which issued sulfurous vapors. One disrobed, sat in the box, and closed a cover that left only the head and neck protruding. "In this horrid contrivance," wrote Campbell, "you can either be skinned alive before you are aware of the danger or else you can be steamed into a damp pulpy condition."

But the food and accommodations were unexpectedly good. "All is comfortable—and unexpectedly so—as a solitary house situated four thousand feet above the sea, on the brink of an always active volcano, on a plain constantly shaken by earthquakes, most of whose luxuries have to come on mule back thirty miles from Hilo over a ragged lava track, could possibly be."

Late in the afternoon they descended 600 feet into Kilauea crater so that they could observe it both in daylight and as night came on. They made their way across a cinder field, past tormented blackened spires of lava—"earth's vomit" Campbell called it—while all around them steam and smoke issued from cracks in the ground. As the sun disappeared behind the rim of the crater they found themselves standing on the edge of a low cliff, watching as the dominant colors shifted from black, gray, and white until a dim crimson glow suffused the drifting clouds of smoke. Beneath them they saw delicate traceries of red and realized that only inches below their feet the molten lava of Hawaii ran in its broken arteries.

HOT SPOT

In the 1960s the new priesthood of the plate tectonic revolution had to deal with rather an embarrassing problem. They had explained how new crust is formed at the mid-ocean ridges and also how it is consumed at subduction zones. But how could they explain the volcanic activity in the middle of the tectonic plates, far from the regions where crust was created or destroyed? Nowhere is this process more active than in the Hawaiian Archipelago and nowhere is the problem more obvious, because Hawaii is more than 3,200 kilometers from the nearest plate boundary. Where do its conspicuous volcanoes come from?

The answer was provided in 1963 by the Canadian geophysicist J. Tuzo Wilson. Wilson was professor of geophysics at the University of Toronto from 1946 to 1974 and one of the architects of the plate tectonic revolution. He is perhaps most famous for explaining the long ridges that occur at right angles to the seafloor spreading centers. He named them "transform faults" and suggested that they are caused by different chunks of crust sliding past each other at different rates as they are formed at the spreading centers. But easily as important was his explanation of volcanoes in the middle of plates. Wilson reasoned that volcanic islands far from spreading centers or subduction zones could form only if there was a localized region where molten magma from the earth's interior welled up and heated the underside of tectonic plates forming localized "hot spots." These hot spots would take the form of volcanoes on the surface. The idea was so radical that when Wilson put it forward it was rejected by all the major scientific journals and was eventually published in the relatively obscure *Canadian Journal of Physics*. In many ways Wilson's experience was a reprise of Wegener's difficulty in getting his theory of continental drift accepted several decades before.

But Wilson's idea of hot spots elegantly explained the linear path of the Hawaiian Island chain from Kauai in the northwest to Hawaii in the southeast and also presented a testable hypothesis, always a sign of good science. The most northerly of the Hawaiian Islands, Kauai, should be both the oldest and the most heavily eroded, while Hawaii itself should be the youngest and least eroded. Radiometric dating of the rocks, as well as the observed degree of erosion on the islands, agreed with Wilson's hypothesis, so the scientific community had no choice but to accept that he was right.

Further research has shown that the Hawaiian Island chain is but the most recent spoor left by the passage of the Pacific plate over this mid-Pacific hot spot. Examination of Heezen and Tharp's map of the northern Pacific shows clearly a submerged line of extinct volcanoes marching northwestward through Midway and beyond. Strangely though, at a latitude of approximately 32° N and a longitude of 171° W, the line of volcanoes kinks abruptly and heads almost due north in the form of the Emperor Seamount chain. This suggests that the Pacific plate changed direction abruptly about 43 million years ago.

As the Pacific plate continues to move northwestward, the active volcanoes of the big island of Hawaii that so enthralled those aboard *Challenger* will gradually become quiescent. In fact, there is evidence to suggest that this is already happening. Thirty-five kilometers to the southeast of the big island the seamount known as Loihi is busily being formed. Loihi is an active submarine volcano and has already risen 3 kilometers from the seafloor to within 1 kilometer of the surface. According to Wilson's theory, Loihi will continue to rise and will one day become either part of the big island or the latest in the long chain of seamounts that extends all the way to the Aleutian Trench.

PACIFIC EXEAT

On August 19, 1875, *Challenger* weighed anchor and left Hilo, heading south for Tahiti and the Society Islands. It was another long dull passage, a full month at sea, because they dawdled as always to sound and dredge despite the growing desire of nearly all on board to get home. It was also on this leg of the voyage that tragedy struck the Scientifics. Von Willemoes Suhm, the young man whom Wyville Thomson recruited in Edinburgh and whose membership of the Scientifics was endorsed by Thomas Henry Huxley, died suddenly of erysipelas, a bacterial infection of the skin. He was only 28. Moseley was devastated:

> I sat with him during the whole of the *Challenger* voyage, working day after day with the microscope at the same table. I am very greatly indebted to him for information in all branches of zoology, and especially in the matter of zoological literature, of which he had the most comprehensive knowledge. I also learnt very much from him in the way of method, and I feel that I shall always remain indebted to him for a decided push on in my general scientific training.

> He was a most indefatigable worker. He was full of hope for the future, and, no doubt, could he have published his journal himself, would have established a reputation as a man of science, which would have been far greater than that which he most deservedly possessed at the time of his death.

With great sadness they buried him in 2,700 fathoms of water, some 300 miles from Tahiti.

It was on this leg of the voyage that they encountered the largest field of manganese nodules they had yet chanced across. On September 16, they brought up more than half a ton and filled two small casks with them. With an average diameter of three-quarters of an inch they resembled nothing so much as the marbles popular with children back home.

Arriving at Tahiti on September 18, 1875, *Challenger* anchored in the pretty harbor of Papeete. All on board were much taken with the beauty of the place, quite in contrast to the dismal scenery they had encountered in the Sandwich Islands. However, they were less impressed with Tahiti's French colonial government. Campbell wrote, "The whole history of the manner in which the French came to occupy this island is irritable and lamentable. Although we may have occupied countries in a high handed manner as regards the natives, still we invariably have something to show for it besides the mere advantages of a naval station, whereas here the French have nothing to show worthy the name of a European power, and this is not because they don't try, but because they do try and fail, which, in two words, is the history of all their colonial attempts."

The French notwithstanding, there was something about the atmosphere Tahiti generated that was hard to put into words, but Joseph Matkin tried.

> This island and people have caused more desertions and punishments in the British Navy than all the rest of the islands of the Pacific put together. Captain Cook lost several men here for some weeks and only recovered them by a stratagem. It was this island and the pleasant memories it excited among the harassed and oppressed crew of the *Bounty* that caused the ship's crew to mutiny near Tongatabu in 1789 ... the longer we stay the more we seem to like it. ... There are five men now in irons for swimming on shore and remaining all night: one man has deserted altogether and there are lots of leave breakers to be punished when we get to sea.

When the time came to leave the beautiful island, they were sorry to go, but Captain Thomson might have been glad to remove his crew from the earthly temptations of plentiful women, food, and drink. They had been at sea for almost three years and all were more than ready to go home. However, there was one reason for relief. Joe Matkin's fears about the tyrannical reputation of Captain Thomson had not been borne out. The new captain was proving himself to be as sympathetic and capable as his illustrious predecessor, George Nares.

They left Tahiti on October 3, played out of harbor by the local band. Under a favorable breeze they shaped a course southeast towards Valparaiso. At first the weather looked set fair for a quick passage and Spry wrote, "all seemed to promise a speedy run over the solitary waste of waters intervening in 5000 miles between Tahiti and Valparaiso." But within a few days they found themselves becalmed and tempers once again became short. It was a great relief then, after six weeks of calm, to sight the island of Juan Fernandez lying in the immensity of the Pacific some 360 miles west of the South American mainland.

The crew's and the Scientifics' interest in the island was more than merely academic though, and more than just a relief from the monotony of the voyage. Juan Fernandez had inspired the setting of the greatest of all castaway yarns: *Robinson Crusoe*. Moseley was fascinated and wrote, "It was with the liveliest interest that we approached the scene of Alexander Selkirk's life of seclusion and hardship, and an island with the existence of which, in the case of most of us, the very fact that we were at sea on a long voyage was more or less distantly connected. The study of *Robinson Crusoe* certainly first gave me a desire to go to sea and 'Darwin's Journal' settled the manner."

The basis for the story is that in February 1704, William Dampier, a noted British buccaneer and navigator, arrived at Juan Fernandez with two ships, both licensed privateers. The second ship was commanded by a Captain Stradling, who quarreled with his shipmates, most of whom demanded to be put ashore. Dampier intervened and his diplomacy eased the situation and, with all but five men left behind, set off into the immensity of the Pacific with the object of some lively maritime acquisition. Some months later,

Stradling returned to the island to find that three of those left behind had been captured by the French. This time the volatile Stradling quarreled with his sailing master, one Alexander Selkirk, so grievously that Selkirk demanded to be put ashore on the island rather than continue to serve with such an intolerable commander. Stradling left Selkirk well provisioned and set sail, ignoring entreaties from the castaway, who changed his mind at the last minute. Selkirk remained on Juan Fernandez until 1709 when he was picked up by *Duke* under the command of Captain Woodes Rogers. Rogers' account of Selkirk's hardship inspired Defoe, although he changed the location to the Caribbean.

Those aboard *Challenger* found the famous island beautiful, its dark basaltic cliffs contrasting with the bright yellow-green of the abundant vegetation, predominantly ferns. Hummingbirds were plentiful and hovered at every bush. A monument to Selkirk had been placed at the crest of a gap between the mountains, 8,000 feet above sea level. There he had sat and watched the sea on both sides of the island in the long-deferred hope of sighting a sail.

On November 15, 1875, *Challenger* said farewell to Juan Fernandez and set sail for Valparaiso, 360 miles distant. That day Captain Thomson issued wine to all hands to commemorate the third anniversary of their commission. They arrived in Valparaiso on December 7th and found it a marked contrast to the beauty of Tahiti. "How Valparaiso came to be called the Vale of Paradise I cannot well understand," wrote Moseley ". . . The surrounding country has a most barren and inhospitable appearance, the red decomposed granite soil showing bare everywhere, and being only here and there sprinkled over with scanty bushes. Not a tree is to be seen anywhere from the anchorage in the harbor, though a wide view is thence obtained of the coast of the Bay."

However, the city was large, prosperous, and very expensive. Matkin wrote, "I think it is the most expensive place that we have been to, a dollar won't go as far as a shilling in England. . . ." By now Matkin, despite his sunny disposition, was feeling, like the rest of the crew, that it was time to go home and get on with his life. Of one thing though he was sure: he would not be staying on in the navy.

Another two members of the expedition *were* staying in the navy but would leave the ship at Valparaiso: Sub-lieutenant A. F. Balfour and the literary bon vivant Sub-lieutenant Lord George Campbell. Both received orders informing them that they had been promoted to the rank of lieutenant. They left *Challenger* in early December 1875 determined to travel to Montevideo, on the other side of the continent, by land so that they could experience the wonder of the Andes and the pampas at first hand. Campbell wrote, "It had, as you know, long been a dream of mine (should I leave the *Challenger* at Valparaiso) to go home overland—across the Andes and the Pampas; and for once in a way my dream came to pass."

Challenger left Valparaiso on December 11th, steaming out of the harbor at daybreak, and then pausing to calibrate the compass. She spent the last days of 1875 and the first days of 1876 surveying in the narrow coves and fjords of South America around the peninsula of Tres Montes, then further south in the Messier and Sarmiento channels before steaming down the Strait of Magellan to avoid the notoriously bad weather of the west Patagonian coast. All around them rose the Andes, the southernmost tip of the Ring of Fire that they had first encountered so long ago on the other side of the Pacific. They had been five months in the Pacific, traveled more than 12,000 miles and explored more than 62 oceanographic stations. On January 20, 1876, *Challenger* left Elizabeth Island in the midst of Broad Reach, the channel that separates "the dreary, barren, mountainous islands of Tierra del Fuego" (as Joe Matkin put it) from the rest of South America and headed for the Falklands. They arrived at Port Stanley, the first British colony they had seen for many months, on January 23, 1876.

They were delighted to be back on British soil once again. "We have never lived better since we left Britain than at this place," wrote Matkin, while Spry commented wryly, "Draught ale and porter at threepence the glass, and good spirits at about English prices, add to the sense of enjoyment of the ship's company." But the easy availability of cheap alcohol was to end in tragedy. After a particularly rumbustuous night of drinking in Stanley a young seaman named Thomas Bush fell overboard. One of the lieutenants, Carpenter, dived in after him and hauled him out but, wrote Matkin of the ship's surgeons, "after persevering for upwards of an hour in the usual methods of resuscitation . . . they were compelled to decide that life was extinct." It was one of the gloomiest days yet for those aboard because the deceased was both popular and engaged to be married on his return.

At Montevideo some days later, there was further sadness at the loss of another literary officer. Herbert Swire had contracted a sickness in Tahiti and was dangerously ill. He was invalided off the ship and sent to catch the first mail steamer home. Matkin wrote, "Sub-Lieutenant Swire was the tallest, strongest, and finest looking man we had in the ship, but now, through bad surgical treatment, he is physically ruined."

From Montevideo *Challenger* headed north for Ascension Island, and then the Cape Verde Islands, where they arrived on April 16, 1876. They were following the line of the mid-Atlantic Ridge, first discovered by them three years before, and by mid-May were passing to the west of the Azores homeward bound. They were not far from a place that would, a hundred years later, see spectacular confirmation of *Challenger*'s discovery.

THE FAMOUS VENT

Bruce Heezen and Marie Tharp used remote-sensing technology to make their earth-shattering discovery of seafloor spreading. But by the early 1970s scientists would settle for nothing less than actually seeing for themselves the place where new Earth was made. By this time the French and the Americans had developed *Trieste* into the undisputed world leader in deep submergence technology. It was this shared expertise that, in 1971, led the French geophysicist Xavier Le Pichon to visit the Woods Hole Oceanographic Institution (WHOI) with a proposition. He suggested to the institution's director, K. O. Emery, that the two nations pool resources and work together to make the long awaited journey to the mid-Atlantic Ridge. Emery's response was positive and the two scientists started soliciting ideas from submergence experts on how to structure and fund the project. And so was born Project FAMOUS (French-American Mid-Ocean Undersea Studies).

One of the people they approached was Bob Ballard, a young scientist who was interested in manned deep-sea diving since his childhood days in San Diego, where he first encountered *Trieste*. Ballard was working with the newly developed American submersible *Alvin* in the Gulf of Maine and thought it could do the job. Le Pichon and Jim Heirtzler, chairman of the WHOI department of geology and geophysics, were named project leaders. Both were strong supporters of the plate tectonic theory and both had trained at Lamont under Maurice Ewing.

The area that the FAMOUS researchers selected for their investigations lay between 36° and 37° N in the middle of the Atlantic Ocean, some 400 miles southwest of the Azores. *Challenger* had passed nearby as she returned to Britain almost exactly a hundred years before. The region was selected because it was known from surface soundings to be seismically active and was considered typical of mid-ocean ridges.

Yet before anyone could go anywhere near the mid-Atlantic Ridge there was much preliminary work to be done. The planning sessions and simulations in many ways echoed those that had recently been so successful in putting a man on the moon. The project leaders managed to persuade the U.S. Navy to let them use its recently developed and highly secret SASS (Sing-Around-Sonar-System) technique to map the study area. SASS was an amazingly powerful tool but, because of Cold War paranoia, had never before been available to the international scientific community. Even after they agreed, the navy brass were too nervous to allow the scientists personal access to the technology itself, so it was naval technicians who surveyed the area and produced topographic maps of extraordinarily good resolution.

The maps were accurate to the nearest 5 fathoms, yet all involved knew that the navy routinely supplied even more detailed maps to its submarine captains for their cat-and-mouse game with the Soviets beneath the Atlantic. The strata underneath the seafloor were mapped and logged using Ewing's subterranean seismic-surveying technique while seismometers placed on the ocean floor around the ridge measured Earth tremors and showed that the sub-sea rift valley was in a state of eternal flux.

The FAMOUS scientists made precise measurements of the heat emanating from the ridge, which confirmed that in this area molten magma from the Earth's interior was welling up into the ocean. In the words of Bob Ballard, this area was the Earth's "crucible of creation." The hopeful sub-sea field geologists also visited the handful of areas above water where they could see the types of features that they were hoping to meet at a depth of 2 kilometers: the volcanic islands of Hawaii and Surtsey and the one land area where a spreading center could be observed directly, the Assal Rift in the East African Rift Valley.

Before they began manned dives, the researchers sent down robotic submersibles, including the LIBEC (LIght-BEhind-the-Camera) system developed by the U.S. Naval Research Laboratory, and WHOI's own ANGUS (Acoustic Navigated Geological Under-sea Surveyor). The landscape they found was anything but silent. It was an alien, inhospitable, volcanic place. Great slumps of suddenly cooled lava made pillow-like formations that clung precariously to the side of the ridge. The ridge itself was a sharply defined hill that ran north and south out of sight into the endless night. There was danger too, such as *Challenger's* Scientifics could never have imagined. Sharp extrusions of lava were like saw blades ready to slice any unwary aquanaut to ribbons. The remote vehicles themselves returned to the surface bruised and dented, powerful witness to the sharp volcanic teeth that awaited the first manned divers.

Xavier Le Pichon himself made the first dive in the French bathyscaphe *Archimede*. He and his crew of two spent much of the time fighting bottom-water currents that almost exceeded the maneuvering ability of the craft. But they were the first to make a remarkable observation that would be confirmed on every one of the 17 other dives made that summer of 1973: The actual zone of crustal formation, that narrow shadow of the tectonic knife, was only a few meters across! This zone, which separated the two huge tectonic plates of Eurasia and America, was little more than the width of a small room.

The following summer the expedition reconvened off the Azores. By now the *Alvin* had been modified to reach the depths required to explore the mid-ocean ridge, using a surplus titanium sphere that WHOI had acquired from the Naval Applied Science Laboratory at the Brooklyn Naval Yards. The previous sphere had been safe only to a depth of 6,000 feet (a little more than a mile) and the mid-Atlantic Ridge lay 9,000 to 10,000 feet below the surface. *Alvin's* superiority over *Archimede* in maneuverability could not be properly exploited until the pressure sphere had been changed.

The advantage of titanium was its strength: Weight for weight it was twice as strong as the stainless steel of *Alvin's* existing hull. Titanium's strength and lightness are what makes it the preferred material in the construction of jet fighters. However, it is hard to work and this caused persistent problems in fitting the penetrators, the openings that allowed vital services like communications and electricity to enter the crew chamber. This delayed *Alvin's* deployment on the mid-Atlantic Ridge until the summer of 1974.

The 1974 dive series, when it did start, was far more ambitious than that of 1973. No fewer than three deep-sea vehicles—the newly fitted out *Alvin*, the French submersible *Cyana*, and the bathyscaphe *Archimede*—began diving early that summer to conduct a coordinated survey of the mid-Atlantic Ridge. Special sequencing of the sonar transmitters on the vehicles and on the surrounding ocean floor was required so that their outputs would not interfere with each other and the three support ships on the surface could track them. Pinpoint accuracy was required to build up a proper set of maps and photographic images of the ridge. Together they made more than 40 dives that summer, retrieving 3,000 pounds of rock as well as dozens of water and sediment samples. In addition the expedition took more than 100,000 photographs.

In this series too they discovered that a much broader zone of tectonic activity flanked the narrow region where the molten magma welled up into the ocean. This process was more or less continuous while the processes that formed the cratered surrounding zone—earthquakes and faulting—were more episodic. The results of the expeditions were written up in an epic two-volume monograph published by the Geological Society of America.

Despite the success of the overall expedition it was clear to all involved that the more maneuverable submersibles *Alvin* and *Cyana* had outperformed *Archimede* and it was this that sealed the bathyscaphes' fate. The French were the first to admit that their *Archimede* was simply too bulky, unwieldy, and expensive, and they retired it. By the end of the 1970s the U.S. Navy had made a similar decision about the world's only remaining bathyscaphe, the *Trieste II*, and so it, too, was mothballed. Small submersibles like the *Alvin* and the *Cyana* had come of age. The availability of lightweight titanium pressure spheres meant that the huge gasoline floats of the bathyscaphes were no longer required and the deep-sea maneuverability that had so long been dreamed off was finally and routinely at hand.

These highly maneuverable deep-sea submersibles would be instrumental in one of the most staggering discoveries of twentieth century marine biology—the underwater volcanic vents where fluids and gases from the Earth's interior blast into the sea, superheating the water around them. Because of the characteristic color of the venting gases, predominantly metal sulfides, these vents were quickly dubbed "black smokers." Black smokers are common in tectonically active areas such as the mid-ocean ridges. Investigations of the black smokers by submersibles soon turned up strange forms of life. These included brightly colored "tubeworms" several meters long, together with clams and other mollusks.

One of the strangest features of the tubeworms is that they have no gut, but have instead a spongy tissue-filled structure known as a trophosome. For a long time it was a mystery how these creatures fed, but the answer came from the materials spewed into the water from the black smokers. These regions are rich in minerals from inside the Earth and support colonies of sulfur-oxidizing bacteria that thrive even at these extreme temperatures. The tube worms and other gutless animals live in symbiosis with these bacteria and exploit their ability to chemosynthesize, that is, to make carbohydrate by breaking down other molecules. It is a perfect partnership, these deep-sea organisms can live far from daylight and exploit a chemical food chain that is analogous to the photosynthesis that powers the sunlit world above. These high-temperature-loving organisms, called "extremophiles," also have commercial uses. They are used to provide enzymes that can work at elevated temperatures and fuel faster chemical reactions in laboratories.

The further exploration of *Challenger*'s first great discovery—the mid-Atlantic Ridge—reached an important milestone when humans eventually visited that dreadful region to see it for themselves, but we should remember that there was an additional impetus for that effort. This came from another science program, another dream of "big science," that also had its origins in the plate tectonic revolution. It was a scientific program that is arguably even more important, if possibly less glamorous, than Project FAMOUS—a scientific program that had the most mysterious of origins in the paranoia of the Cold War, the space race, and the missile gap.

BLACK OPS AND COLD-WAR ROCKS

For Harry Hess, as for many career scientists, the price of fame and success in science was a crushing burden of administrative duties, both in his home institution as the head of a department and also externally, in serving on committees. One of the committees Hess served on in the 1950s was an evaluation committee for the National Science Foundation, the branch of the U.S. government that doles out money for scientific research. The function of the committee was to determine the best proposals put forward that year by geologists around the country and recommend to the NSF those worthy of funding.

In July of 1957, just six months before Sputnik was launched, the committee met to evaluate a bunch of geological and geo-physical proposals and concluded that many were laudable but none were earth shattering. This conclusion generated a feeling among the committee that the earth sciences needed their own big-science equivalent to developments in astronomy, which was in the midst of a growth spurt fuelled by the space race. Also present at that funding meeting was Walter Munk of the Scripps Institution of Oceanography in California. Half in jest, Munk offered that, in the absence of anything else to fund, perhaps they should themselves put forward a proposal to drill through the Earth to the Mohorovicic Discontinuity.

This was discovered in 1909 by the Croatian geologist, Andres Mohorovicic, who had been measuring the time seismic waves take to reach different seismometers. The waves' speed of propagation was known to be on the order of 5-6 kilometers a second but, to his surprise, Mohorovicic discovered that certain waves, those that took a deep routing through the interior of the Earth, appeared to reach seismic stations on the part of the Earth's curve opposite their point of origin faster than those waves that traveled shallow. Mohorovicic concluded that there was a layer deep in the interior of the Earth below which seismic waves were accelerated. This layer became known as the Mohorovicic Discontinuity or Moho. There was much debate about the nature of the Moho layer and it was this that sparked Munk's suggestion. Hess suggested that they refer the project to the secretive and little known American Miscellaneous Society for action.

The American Miscellaneous Society (AMSOC) was formed in the summer of 1952 by Gordon Lill and Carl Alexis, two geologists working at the Office of Naval Research (ONR), to consider ideas that did not fit into established categories of thinking. It was deliberately loosely structured and had no statutes or formal membership, no minutes, no staff, and no organizational chart. It existed solely to *get things done* and commonly met in the rooms of the Cosmos Club in Washington, D.C. where outré ideas were booted around freely. AMSOC was very unofficial, yet it attracted some of the best minds in America.

There seems to have been a kind of intellectual optimism in the air that had its roots in the scientific successes of the Second World War. Science was fun and the people who did it were determined to have fun, too. There were even divisions of AMSOC named "Committee for Co-operation with Visitors from Outer Space" and the "Society for Informing Animals of their Taxonomic Positions." AMSOC's power lay in its connection to ONR.

President Harry Truman formed ONR in 1946 to "plan, foster and encourage scientific research in recognition of its primary importance as related to the maintenance of future naval power and the preservation of national security." The ONR was charged with establishing research programs in collaboration with universities and corporate laboratories across the length and breadth of the United States. Although the programs did not need to have practical applications to be approved, those that did would be able to apply for development funding, especially if they looked likely to have a military use. It is impossible to overstate the importance of the ONR in the postwar years and its influence in funding science because it underwrote so many fundamental scientific projects.

These projects included the Electronic Numerical Integrator And Computer (ENIAC), the first computer ("small enough to fit into a room"); the development of superconductivity, turbojets and gas turbines; and also, as we have seen, the development of deep-submergence technology in the form of the *Trieste II* and the *Alvin*.

On April 20, 1957, Walter Munk hosted a breakfast meeting of AMSOC at his home near Scripps to discuss the Moho proposal. AMSOC agreed to pursue it and called the project the Mohole. They knew already that land-based drilling would not allow them to reach the Moho. The thickness of the continental crust, some 30 to 50 kilometers on land, would wear out any drill bit through heat and friction long before it got anywhere close to the enigmatic boundary layer. Drilling from the ocean, though, was a different matter. Geophysical surveys had already shown that the thickness of oceanic crust was only about 5 kilometers, an amount they estimated as relatively easy to penetrate if only they could get the drill bit positioned correctly under water.

On April 27, the Mohole committee reconvened at the Cosmos Club. Also in attendance that day was Maurice Ewing. Since 1953, Ewing had been traveling America trying to get people interested in deep-sea drilling, not to reach the Moho, but rather to retrieve sedimentary sections that would illuminate the history of the oceans. By September 1957 the Mohole committee had acquired more members from the geological community and met to discuss the project in open debate at a meeting of the International Union of Geodesy and Geophysics in Toronto, Canada. In an eerie harbinger of the Soviet space spectacular that within weeks would spur America to commit to the goal of putting a man into space, a Soviet scientist stood and said that the U.S.S.R. already had advanced plans to drill to the Moho. To guard against supposed Soviet espionage, at the next meeting of the committee armed guards attended to ensure the security of the deliberations. Such was the paranoia of Cold War America.

In conjunction with John Mecom, a deep-drilling oil specialist, the committee formulated a plan to reach the Mohole. There would be three phases: Phase One, a practice hole drilled on near-shore continental crust; Phase Two, a sedimentary section hole drilled in the deep ocean; and Phase Three, the final push, a hole drilled in the deep ocean to reach the Moho.

In early 1958 the AMSOC Mohole committee approached the NSF for initial funding of $15,000 to conduct a feasibility study. NSF declined to fund an enterprise conducted by such an informally structured organization, so Harry Hess asked that the AMSOC Moho committee be allowed to formally associate itself with the National Academy of Sciences (NAS). Hess was already a fellow of the NAS and his proposal was well received. The physicist I. I. Rabi, one of the Nobel laureates on the committee that heard Hess's request, remarked dryly, "Thank God we're finally talking about something beside space." It was an ideal association for the AMSOC committee. The NAS dates back to Lincoln's time and is nothing if not respectable. It was and is the most distinguished scientific body in America, equivalent to Great Britain's Royal Society, which had underwritten the original *Challenger* expedition.

At the American Geophysical Union meeting in Toronto that year it was agreed that the drilling barge *CUSS I* would be used for Phase One. *CUSS I* was specially constructed by the Global Marine Exploration Company of Los Angles, California, and paid for by a secret consortium of four oil companies: Continental, Union, Shell, and Superior. Engineer A. J. Field showed a movie of the vessel, carrying a full-sized rig, drilling an experimental well off the California coast in 200 feet of water. Until that moment the existence of mobile oil platforms had been kept a closely guarded commercial secret. No one present had even heard of such equipment. But all could see that the technology to reach the Moho was, after all, within reach.

Now that the Mohole committee was affiliated with the NAS and had the backing of the entire American geophysical community, the NSF immediately provided the cash for a feasibility study of possible sites in the Atlantic and the Pacific. Rumors were beginning to fly in the press about the project: The hole would have to be 10 miles deep. The rock at the bottom would be too hot to permit conventional drilling. Secret and experimental technology was being devised to do the job. The cost of the project would be on the order of the space program. There was a whiff of terror in the air, too: What if the Mohole scientists destroyed the world with their experiment? That fear inspired a major film studio, Paramount, to make the disaster movie, *Crack in the World*, in which a deep-drilling project causes a huge crack to grow and grow until it threatens to split the earth in half. To scotch such rumors, the AMSOC committee decided to go public. In April 1959 they published an article in *Scientific American* outlining the goals and technologies of the project. It helped, some.

In late 1960 a contract was drawn up between the NSF and Global Marine to modify the *CUSS I* for truly deep drilling. Two technical solutions were needed immediately: a way to keep the drill string vertical above the drill hole and a way to keep the vertical motion of a floating vessel from smashing the drill bit to pieces.

At that time the Bendix-Pacific Corporation was heavily involved in providing high-technology solutions for the American space program. Because the Mohole project was being billed by Munk and his associates as the earth sciences' answer to the space program, it was no surprise that Bendix, in collaboration with the Mohole project director, Willard Bascom, provided the answer to the first problem. Bascom and Bendix devised what they called a dynamic positioning system, whereby the drilling barge was ringed by a system of six "taut-line" buoys. A sonar transponder on board the vessel bounced signals off the buoys and a computer used this information to keep the vessel centered in the middle of the ring, using a series of propellers spaced around the hull, holding it stationary within a circle of 50 yards.

The second problem, damping the vertical motion of the drill string, was solved by the invention of bumper subs. Effectively, this meant disconnecting the drill string and enclosing it in an external sleeve that could absorb the vertical motion of the ship. The sleeve was then attached to the drill string by a set of gears to transmit rotary motion.

CUSS I drilled the first hole near Scripps at the beginning of March 1961. The drill bit penetrated 315 meters into the seafloor while the ship floated on almost a kilometer of water. By anybody's reckoning the first test was a success. In April the barge relocated to a site near the island of Guadalupe off the west coast of Mexico and successfully held station for three weeks, drilling five more holes in water as deep as three-and-a-half kilometers. Hundreds of meters of sediment were recovered as well as underlying volcanic rock. It was clear that the combined forces of AMSOC and the NAS did indeed have the technology. An exultant President Kennedy, already flushed with the success of Alan Shepard's sub-orbital flight in May of that year, sent the barge a congratulatory telegram praising it as a "historic landmark."

Yet there was another problem that the Mohole Project faced, and it appeared insurmountable. Everyone knew that the drill bit would have to be changed—probably several times—before the Moho could be reached, and for that the drill string would have to be pulled out of the hole. Just how were they going to get the string back into the same, tiny hole in the seafloor having lowered it again all the way from the ship?

It would be 10 years before that problem was solved, by a successor to the Mohole Project. The cost of drilling the Guadalupe experiment had been in the region of $1.8 million and initial estimates for the cost of the second phase were around $15-20 million. By 1965 that cost had escalated to a staggering $112 million and yet the project seemed to have made little progress. With technical problems and cost overruns dogging the program in August 1966, Congress cancelled the Mohole Project.

But every cloud has a silver lining. CUSS I's drilling proved that deep-sea drilling was feasible and that if enough money were spent there were no technical problems that could not be overcome.

At the University of Miami a young geologist, recently hired, had been for some time very interested in another problem of the deep sea. For Cesare Emiliani the emphasis was not on the hard rocks that underlay the deep ocean's sediments but on the sediments themselves. Emiliani was convinced that the way to calculate the number of ice ages in the recent past was to look at deep-sea sediments. He reasoned that if the ocean floor was really a silent landscape where sediments accumulated uniformly and undisturbed across millennia, surely it was here that the best evidence for the sequence of the ice ages would be found. The technique that he would use to calculate the number of ice ages was the oxygen-isotope technique of Harold Urey that he had learned at the feet of the master himself.

Emiliani was appointed to the staff of the University of Miami in Florida on January 1, 1957. His appointment there was strongly supported by faculty member Bob Ginsburg, the famous Caribbean coral scientist, who is legendary for his support of young geologists. Legend has it that one day in early 1962 Ginsburg passed Emiliani on the way back from lunch. He enquired how the new faculty member was getting on and they started chatting. Ginsburg listened sympathetically to Emiliani's thoughts about the applications of oxygen isotopes to deep-sea cores and then said simply, "You should be thinking big." Emiliani asked him what he meant. "Just think big," said Bob and walked away.

As we have seen, at that time the oxygen-isotope technique of temperature measurement was producing estimates of the number of glacial stages that were quite at variance with the number suggested by those who counted fossils and Emiliani was desperate to defend the oxygen-isotope technique. He realized that his problems would be largely solved if he could get more, and especially deeper, holes—holes that contained more of the Pleistocene. On top of this was the fact that he had long been privately critical of the money being spent on the Mohole Project.

The lure of simultaneously vanquishing his bug-counting enemies and spiking the Mohole Project's guns was too tempting to resist. He went back to his office and immediately telephoned R. F. Bauer, the president of the Global Marine Company of Los Angeles and enquired about the feasibility of drilling two holes in the deep water south of Puerto Rico. He was told that it could be done for $2 million. Immediately he wrote to John Lyman, program director of oceanography at the NSF, and asked for the money. To his amazement he was encouraged to put his proposal in writing, which he did in late March 1962. By June the first meeting of the so-called LOCO (LOng COres) committee (sponsored by Lyman under the auspices of the NSF) was convened in Miami.

Not only was Emiliani's proposal endorsed, but the committee also recommended that more money be made available to drill more holes farther out in the Atlantic. Emiliani was elected chairman of the new committee and by September LOCO had agreed to elect four sponsoring institutions to guide the project. These institutions, the "big four," which together became known by the acronym JOI (Joint Oceanographic Institutions), were WHOI, the University of Miami, Scripps and the Lamont-Doherty Geological Observatory of Columbia University in New York. The presence of Lamont was no big surprise, because Doc Ewing had got into the act early on as a vociferous and very well-connected supporter of Emiliani. For several months there was much maneuvering but finally a consensus emerged. When LOCO became CORE (Consortium for Ocean Research and Exploration), the original four were joined by the University of Washington (giving representation to an institution from the Pacific Northwest for political reasons) and the organization spawned JOIDES—Joint Oceanographic Institutions for Deep Earth Sampling.

In April of 1965, when the Mohole Project was attracting enormous publicity and beginning to generate some criticism, a drilling vessel on loan to JOIDES known as *Caldrill I* put down several test holes in the Atlantic. The results were spectacularly successful and confirmed the viability of deep-sea drilling into sediments. Even as the Mohole Project died in 1966, a contract to run the newly formed Deep Sea Drilling Project (DSDP) was awarded to one of the founding JOIDES members, Scripps. A new drilling vessel was to be built especially for the enterprise by the same people who constructed the *CUSS I* and the *GLOMAR Explorer*, Global Marine. In honor of the expedition of 1872–1876 that founded the sciences of oceanography and marine geology the drilling vessel was named *GLOMAR Challenger*.

The keel of the *GLOMAR Challenger* was laid on October 18, 1967 in Orange, Texas, and the ship was launched on March 23, 1968 from that city. It sailed down the Sabine river to the Gulf of Mexico, where it underwent several months of testing before being accepted by the DSDP, the operational arm of JOIDES, on August 11, 1968. The voyages of the *GLOMAR Challenger* were named legs— the idea being that the legs would eventually follow one another seamlessly, without any time gaps to maximize the NSF's invest-ment, and keep the ship at sea as much as possible.

The *GLOMAR Challenger* was 400 feet long, displaced 6,281 tons, and supported a 45-meter derrick that could lift a drill string weighing one million pounds, equivalent to a pipe length of 7 kilometers. The ship was completely self-sustaining, carrying enough food, water, and fuel to remain at sea for up to 90 days at a stretch. In the center of the ship was the moon-pool, a circular opening directly beneath the derrick through which the drillstring was lowered into the water.

The vessel carried a crew of about 70 including the ship's officers, drilling roughnecks and technicians, as well as the co-chiefs and staff scientists selected for the voyage by the DSDP. The scien-tific complement numbered between 9 and 16, depending on the length and complexity of the leg and they, like the co-chiefs, were divided into two shifts. When the ship arrived on station it was held in position by dynamic positioning thrusters based on the design used aboard *CUSS I*. The whole operation eventually became so streamlined that two samples (long plastic tubes of sediment) per hour could be retrieved. Since its inception, this successor to the original *Challenger* expedition has enjoyed many successes some of which have been mentioned in this book, for example, the evidence, discovered early in Leg 13, of the Mediterranean's desiccated past.

The very first leg of the DSDP was jointly overseen by Doc Ewing and geologist Joe Worzel, another scientist who would eventually make his name explaining the mechanics of continental drift and seafloor spreading. Ewing's selection as the first chief scientist to guide the *GLOMAR Challenger* was a very public recognition of his pre-eminence in the field of oceanography. But the first leg of the project did not investigate the phenomenon of seafloor spreading; that would have to wait until the ship was ready to brave the true ocean environment of the Atlantic. The ship first drilled in the Gulf of Mexico and in the process managed to prove one of the Doc's pet theories: that there were salt domes in the Gulf and that these were the cap rocks for oil reservoirs.

After the Gulf shakedown leg, the ship went to New York where it was prepared for one of its most important tests. Having proved itself in the relatively safe environment of the Gulf, the *GLOMAR Challenger* would sail across the Atlantic to Dakar, Senegal, in Africa, and attempt to bring back evidence to test the hypothesis that now had the geological world agog: continental drift and seafloor spreading.

Leg 2 of the DSDP was designed to test the hypothesis in light of Fred Vine and Drummond Matthews's interpretation of the stripes of reversed and normal polarity that flank each side of the mid-ocean ridges. Assuming that the rate of new seafloor production is more or less constant, the age of successive magnetic stripes on the seafloor would directly translate to distance from the ridge axis. That is, the farther you go from the ridge crest, the older the seafloor you are standing on. This meant that the edges of the ocean basin should have the oldest basement rock and should have had time to accumulate the thickest sediment cover. Early calculations had already suggested that the seafloor was spreading at a rate of 1.25 centimeters per year.

Here, then, was a test of the seafloor-spreading hypothesis that could be done only by a drilling ship. By drilling into these sediments and measuring their ages using the microfossils so lovingly described by the original *Challenger* expedition, it would be possible to prove or disprove whether time did indeed equal distance. Although Leg 2 was dogged with technical problems, it proved beyond doubt that the deepest sediments do indeed become younger as you approach the spreading centers at the mid-ocean ridges. The scientists aboard *GLOMAR Challenger* were even able to calculate the rate of seafloor spreading from their observations: 1.2 centimeters per year, a figure that agreed well with the estimate of 1.25 cm per year. Close enough for government work, as the saying goes, and certainly enough for the *Washington Star* to lead an editorial with the words, "It's further to Europe, study finds. . . ."

On the next leg of the DSDP, the ship returned across the South Atlantic to Rio de Janeiro. This leg was as trouble free as its predecessor was troubled, and the results more than confirmed the data recovered by the *GLOMAR Challenger* in the North Atlantic.

Two voyages of the *GLOMAR Challenger* proved beyond doubt the theories of continental drift and seafloor spreading which combined underpin plate tectonics. Over the years the DSDP produced many other spectacular discoveries until it was replaced in 1983 by an even more ambitious program called the Ocean Drilling Program (ODP). The ODP comes to an end in 2003 but will in turn be replaced by the Integrated Ocean Drilling Program, involving more drilling vessels as well as standalone drilling platforms.

Dreams of big science are still being dreamed, and they are all based on the earliest dream of them all, the voyage of HMS *Challenger* between 1872 and 1876.

ARRIVAL

The weather after the Azores was poor, "with strong and adverse winds," as William Spry put it. It was decided that *Challenger* would put into Vigo (Portugal) to coal and this they did on May 20. They did not linger, because the lure of home was too strong. Early the next day they were again at sea. "The weather was still squally and unpleasant," wrote Spry, "yet we managed to get round Cape Finisterre; and now with the wind somewhat fairer, a capital run was made across the dreaded Bay of Biscay." On the evening of the 23rd they saw the light on Cape Ushant and the next morning gazed through welcome haze and fog at the soft green lines of Old England. "Onward we go, sighting the old familiar headland and landmarks—the Eddystone, the Start, the white cliffs at Portland and St. Alban's head—until at last the Needles are in sight...."

They were home.

Epilogue

The Scientifics began writing the report of the voyage almost as soon as they returned to their home institutions. Their activities were coordinated from the *Challenger* office in Edinburgh, and subspecialists, such as the noted foraminiferal expert, H. B. Brady, were commissioned to write some portions. But not many years went by before a tragedy befell them that dwarfed even the loss of von Willemoes Suhm. After publishing his own account of the voyage (concentrating mostly on the Atlantic), Charles Wyville Thomson, who had overextended himself for the three-and-a-half years that *Challenger* circled the globe, died of exhaustion before the first volume of the official report was published. His place as leader of the Scientifics was taken by John Murray who, after consulting with his colleagues, published the first volume under the author-ship of Staff-Commander T. H. Tizard, RN, Professor H. N. Moseley, FRS, Mr. J.Y. Buchanan, MA, and Mr. John Murray, PhD, Members of the Expedition.

John Murray was a giant who filled a giant's shoes. In the end the *Report*, with all its specialist appendices, ran to 50 volumes, the last published in 1895. By this time the Treasury had wearied of paying for the project and it was Murray who invested a consider-able portion of his fortune in completing its publication. Perhaps he felt that this was an appropriate gesture, because his fortune was a direct result of the expedition. While in the Moluccas he had noticed considerable amounts of phosphate. Returning to the area in later years he founded the Christmas Island Phosphate Company for fertilizer production. This venture became so profitable that before Murray's own death in a motor vehicle accident (near his home Challenger Lodge, not far from Edinburgh, in 1914) the royalties the company paid to the British government had already surpassed all the costs of the *Challenger* expedition.

After *Challenger's* return, Henry Moseley was immediately elected to a fellowship at his old college, Exeter, in Oxford, where he spent the next several years working up his part of the results of the expedition. During this time he wrote *Notes by a Naturalist on* HMS *Challenger*, a charming and sensitive account of the voyage, which was published in 1879 by the venerable London publishing house of John Murray, who had also published Darwin's *Origin of Species by Means of Natural Selection*. Moseley's work on *Peripatus*, performed during the *Challenger* expedition, along with his work on corals, was judged so important that he was awarded the Royal Medal of the Royal Society in 1887, after which he set off to study the anthropology of the North American Indians of California and Oregon. In 1879 he was elected a fellow of the Royal Society, the highest British honor that a scientist can receive. He was appointed Linacre Professor of Zoology in the University of Oxford in 1881 and immediately threw himself into the task of reorganizing the natural history course and establishing the Pitt-Rivers collection of anthropological artifacts.

In later years anthropology was to become Moseley's overriding passion. He worked at a feverish pace and eventually his health suffered. Headaches and depression assailed him and complete collapse followed. For more than four years his devoted wife nursed him but on November 10, 1891, four years before the final *Challenger* report was published, he died of bronchitis at the age of 47. Herbert Swire summed up the man in a letter that he wrote after Moseley's death: "He brought to his investigations ability and perseverance of no ordinary kind, backed by an originality of mind and an imperturbable good humor, which made him absolutely proof against all the shafts with which naval wit was never tired of trying the mettle of those whom we called our philosophers. . . . Personally I always looked on Moseley as one of my greatest friends. . . ."

The expedition's chemist, John Young Buchanan, returned to Edinburgh, where he set up a private laboratory. In later years he had close ties to Christ's College, Cambridge, and often sailed with Prince Albert I of Monaco on his oceanographic cruises. Buchanan's chief contribution to the cruise was the debunking of a theory that had been put forward by the German scientist Ernst Haeckel, namely that the floor of the ocean was covered in a primordial slime that was even awarded a Latin name, *Bathybius huxleyii*. Huxley himself was a great proponent of the existence of this mythical creature but it was Buchanan who proved that it was merely sulfate of lime that precipitated in preserving jars when seawater was mixed with the preserving fluid. On hearing the news, Huxley, with characteristic aplomb remarked wryly, "*Bathybius* has not fulfilled the promise of its youth." Buchanan died at the age of 82, the last survivor of *Challenger*'s civilian scientific staff.

Several of the navy men published accounts of their experiences aboard *Challenger*. To a fascinated Victorian public, certainly the best known of these was William J. J. Spry's account *The Cruise of* HMS *Challenger* published in 1877. It ran to twelve editions, six of them in the first year of publication alone, and was still in print as late as the 1890s.

Lord George Campbell's account, *Log Letters from the Challenger*, published by Macmillan in 1876, remains a fascinating and more personal insight into the rigors of such a long voyage.

The letters that Joseph Matkin wrote home were deposited with the Scripps Institution of Oceanography as late as 1985 and published in 1993. This is the first time they have been used as the basis for an account of the general *and* scientific achievements of the voyage of HMS *Challenger*. As he had promised his mother,

Matkin left the navy and returned home to Rutland before moving to London as a civil servant. In 1880 he married Mary Swift of Oakham who bore him five sons in the 14 years they lived in London. In 1894, the 41-year-old Matkin retired from the civil service and returned with his family to Oakham, where his last son was born. In 1914 he returned to London, having separated from his wife. He died in 1927 from complications following a motor-vehicle accident.

Herbert Swire, despite the illness that assailed him in Tahiti, lived a long and fruitful life. He arrived back at Southampton from South America, having recovered his health during passage aboard the mail steamer *Douro*, on March 18, 1876, some six weeks before *Challenger* herself arrived back at Spithead. He was the last of the *Challenger* men to die, in 1934, and left behind a son who was responsible for the private publication, in 1938, of his beautifully illustrated journal of the voyage.

As we look back from the early twenty-first century, what, then, is the legacy of the *Challenger* expedition? In truth, its importance can hardly be exaggerated. It set the scene for the plate tectonic revolution of the 1960s and 1970s, and its discovery of manganese nodules has enabled the exploitation of a vital new resource on the seabed. The cataloging of seafloor sediments by Murray and Wyville Thomson led the way, in the middle of the twentieth century, to the unraveling of the history of climatic change, so vital for our own future on a warming Earth. The appreciation of the importance of missing links found in the antipodes of the British Empire fuelled twentieth-century biology and set the scene for the so-called "modern synthesis" of evolutionary theory, where paleontology and modern genetics were finally welded into a synergistic whole.

But it is the less tangible aspects of *Challenger*'s legacy that demand attention and respect. It was the first great voyage of scientific exploration, sent out with no other purpose than the acquisition of knowledge. It was a milestone in the history of humanity, when the importance of learning *for its own sake* was perceived, not just by a small intellectual elite, but by ordinary people as well. Joe Matkin's letters show that clearly.

This appreciation of learning answered the God versus Science question too, finally laying to rest the belief that secular questions can be answered by religion. Although the Scientifics, officers, and bluejackets might never have perceived it themselves, in that three-and-a-half-year mission they placed God and the Church in their rightful place: arbiters of the spiritual questions that can never be answered by the methods of science, while showing clearly that scientific questions are properly the province of science.

The most tangible legacy of the voyage of HMS *Challenger* though must surely be the great ocean drilling programs of the late twentieth century—the Deep Sea Drilling Project (DSDP) and Ocean Drilling Program (ODP), whose findings I have mentioned throughout this book. There is no clearer indication of the importance of HMS *Challenger*'s voyage than that the first dedicated scientific drilling ship in history, *GLOMAR Challenger*, was named after her. *GLOMAR Challenger* was retired in the mid-1980s when the ODP replaced the DSDP and a new ship, the *JOIDES Resolution*, was commissioned. Now the *Resolution*, in its turn, is about to hang up its drill string and the ODP is to be replaced by the Integrated Ocean Drilling Program (IODP) with several drilling ships and platforms, some able to drill in oil-rich areas, places where its predecessors could not go.

Finally, we should remember that two of humankind's greatest technical achievements were named after HMS *Challenger* too: the Lunar Module of the *Apollo 17* mission, as well as the space shuttle OV-99 that tragically exploded during launch above Cape Canaveral in January 1986, were named after that same, small, Victorian sail-and-steam corvette.

As we honor the Apollo astronauts as well as the crew who lost their lives aboard the space shuttle *Challenger,* perhaps we should also spare a thought for the ship for which those technological marvels were named, and recall that perilous voyages of discovery have always been a part of our indomitable human spirit.

Acknowledgments

Many thanks to Jeffrey Robbins, Máire Murphy, Ann Merchant, Dick Morris, and Robin Pinnel at the Joseph Henry Press and Grant McIntyre at John Murray for careful editing and excellent suggestions, which have much improved the manuscript. Special thanks to my agents Peter Robinson and Jill Grinberg, whose tireless support has been an inspiration even in the darkest hours.

I have benefited from discussions with many colleagues in Oxford and beyond: thanks to Hugh Jenkyns, John O'Sullivan, and Steve Simpson, and particularly Mike Durkin, Tony Watts, Brent Dalrymple and Marcia McNutt who read the manuscript and made comments. Any omissions or errors of course remain my own. The librarians at various Oxford libraries have, as always, been wonderful, so thank you Stella Brecknell, Theo Dunnet, John Hillsdon, Isabel McMann, Helen Morgans-Bell, Karen Smith, and Rebecca Sparkes. Moral support has, as always, been supplied by many good friends at the Frewen Club, in particular Dennis and Christine Armstrong, Brian Boyt, Tom Duggan, Colin Holmes, and Peter Kent. Thanks also to other friends: Alan Edwards, Marvina Houghton, Peter James, Robert McKenna, Jon and Jo Oldham, and Sonia Newton-Clare.

Thanks to the Society of Authors who supported the writing of this book with a grant from the Kathleen Blundell fund and to Julie, Jessica, and Susannah who put up with Dad in the backroom yet again.

Further Reading

Ballard, R. D., 2000. The Eternal Darkness: A Personal History of Deep-Sea Exploration. Princeton: Princeton University Press.

Bascom, W., 1961. A Hole in the Bottom of the Sea. New York: Doubleday and Co.

Briggs, P., 1972. 200,000,000 Million Years under the Sea. London: Cassell.

Broad, W. J., 1997. The Universe Below: Discovering the Secrets of the Deep Sea. New York: Touchstone Books.

Buchanan, J. Y., Moseley, H. N., Murray, J., and Tizard, T. H., 1895. The Report of the Scientific Results of the Exploring Voyage of HMS *Challenger* during the years 1873-1876. London, Edinburgh, and Dublin: Government Printing Office.

Campbell, Lord George., 1877. Log–Letters from the *Challenger*. London: Macmillan.

Darwin, C., 1845. The Voyage of the Beagle: Journal of Researches into the Natural History and Geology of the Countries Visited During the Voyage of H.M.S. *Beagle* Round the World. London: John Murray. *Also* New York: Modern Library, 2001.

Deacon, M., Rice, T., and Summerhayes, C., 2001. Understanding the Oceans. New York: Routledge.

Gould, S. J., 2000. Wonderful Life: The Burgess Shale and the Nature of History. New York: Norton.

Hsu, K. J., 1983. The Mediterranean was a Desert: A Voyage of the *GLOMAR Challenger*. Princeton: Princeton University Press.

Hsu, K. J., 1992. *Challenger* at Sea. Princeton: Princeton University Press.

Kunzig, R., 2000. Mapping the Deep: The Extraordinary Story of Ocean Science. New York: Norton.

Linklater, E., 1973. The Voyage of the *Challenger*. New York: Doubleday.

Lippsett, L., 1999. Lamont-Doherty Earth Observatory: Twelve perspectives on the first fifty years 1949-1999. New York: Office if Communications and External Relations Lamont-Doherty Earth Observatory.

Menard, H. W., 1986. The Ocean of Truth: A Personal History of Global Tectonics. Princeton: Princeton University Press.

Moseley, H. N., 1892. Notes by a Naturalist: An account of observations made during the voyage of HMS *Challenger* round the world in the years 1872-1876. New York: AMS Press (a new and revised edition).

Rehbock, P. F., 1993. At Sea with the Scientifics: The *Challenger* letters of Joseph Matkin. Hawaii: University of Hawaii Press.

Spry, W. J. J., 1895. The Cruise of HMS *Challenger*. London: Sampson Low, Marston and Company,.

Swire, H., 1937. The Voyage of the *Challenger*. London: Golden Cockerel Press.

Warme, J. E., et al. eds., 1981. The Deep Sea Drilling Project: A Decade of Progress. Tulsa: Society of Economic Paleontologists and Mineralogists Special Publication, 32.

Wyville Thomson, Sir C. 1877. Voyage of the *Challenger*. London: The Atlantic.

Index

A

Admiralty Islands, 203, 222
Africa
    coral reefs, 69
    lungfish, 182
    "White Man's Grave," 67, 112
Afro-European continental shelf, 58
Agonic lines, 75–76
Agulhas Current, 88, 132
Albatross, 148, 149
Albert I, Prince of Monaco, 251
Aldrich, Pelham, 202
Aldrin, Buzz, 209
Aleutian Trench, 226
Alexis, Carl, 238
Algeria, 43
Aluminum, 53
Alvin submersibles, 232, 234–235,
        239
Amboyna, Dutch East Indies, 199
American Geophysical Union, 240
American Miscellaneous Society
        (AMSOC), 238–239, 240,
        241, 242
Amundsen, Roald, 165
Anchialine organisms, 97–98

Ancoma (German frigate), 178
Andes Mountains, 230
Andries Vening Meinesz, Felix, 196
Anemones, 107, 214
ANGUS (Acoustic Navigated
        Geological Undersea
        Surveyor), 233
Anhydrite, 41
Annelids, 187, 191
Antarctic Bottom Water (AABW),
        154, 156, 157
Antarctic Circumpolar Current, 153
Antarctic Convergence, 153, 156
Antarctica, xiii, 94
    biological fecundity of waters, 157
    climatic effects, 154, 156, 158–159
    currents and seas, 153–154, 156,
        157
    desert characteristics, 154
    energy resources, 77
    discovery, 129–130
    Europa analogue, 170–171
    freshwater resources, 154, 168–170
    ice accumulation rates, 161
    ice cores from, 160–162
    icebergs, 154–156, 159

Lake Vostok, 168–170, 173
Maud Rise, 79, 81
precipitation, 154
separation from South America,
    157–158
temperatures, 160
thermal isolation, 158
upwelling zones, 138, 157
Vostok Ice Station, 160–162, 168–
    169
Antrim, 151
Apollo missions, 24, 172, 173, 253
Arafura Sea, 197
Archaean era, 70
Archimede bathyscaphe, 234–235
Arctic Basin, 54, 55
Arctic exploration, 22, 50
Arctic icebergs, 154, 156
Arctic resources, 77
Arcturus (yacht), 206
Armstrong, Neil, 209
Around the World in Eighty Days, xi
Arthropods, 187, 191
Aru island, 198
Ascension Island, 231
Ashanti Wars, 112, 130–131
Assal Rift, 233
Aswan Dam, 43
Atherstone, Dr., 131
Atlantic (ship), 86, 91
Atlantic easterly drift current, 132
Atlantic Ocean
    circulation system, 88–89
    deepest part of, 196
    earthquake epicenters, 59
    mid-Atlantic ridge, 53, 55–60, 62,
        84, 98, 141, 147, 231–236
    topographic charts, 29
    zone of crustal formation, 234

Atlantis (ketch), 141
Atolls, formation of, 71
Atomic bomb testing, 163
Attenborough, David, 69
Australia, 179, 197
    Aboriginals, 184
    Great Barrier Reef, 73, 193
    missing links, 181–184, 187
Aysheaia, 189
Azoic theory, 2–3, 4, 5, 62, 210, 217
Azoic zone, 2
Azores, 108–113, 231, 234, 248

B

Bacon, Francis, 56
Bacteria
    bioluminescence in, 114
    in ice cores, 169–170
    in polymetallic nodules, 54
    sulfur-oxidizing extremophiles,
        236
Bahamas, 89
Bahia, Brazil, 121–126
Bains, Santo, 81, 82
Balfour, A. F., 200, 201, 230
Ballard, Bob, 232, 233
Banda Island, 198–199
Barium, in sediments, 81
Barramundi, 180
Barton, Otis, 207–212, 214
Basalt, 23
Bascom, Willard, 241
Batchian island, 199
Bathybius huxleyei, 251
Bathyscaphes, 205, 212–217, 231,
    234–235
Bathyspheres, 206–209, 212
Bauer, R.F., 244

Bay of Biscay, 21, 22, 248
Beebe, William, 205-212, 213, 214
Belgium, 213
Bellerophon (ironclad), 35
Ben Mor, 185
Bendix-Pacific Corporation, 241
Benthoscope, 212, 214
Bermuda
    Castle Harbor, 97
    climate, 91
    coral reef system, 73, 84
    deepsea manned exploration, 206, 209-212
    freshwater, 95
    Hamilton Harbor, 84-86, 94-99, 100
    Harrington Sound, 95, 97
    naval cemetery, 86, 99
    strategic importance to British, 85
    structure, 67
    Walsingham Caves, 95-99
Bermuda Triangle, 74, 75-76, 78-79, 82, 84
Bimini Island, 88
Bingham Oceanographic Laboratory, 105
Biodiversity hotspots, 97
Bioluminescence, 113-115, 210
Blake Plateau, 78-79, 81, 82
Bluejackets/tars
    burial at sea, 74
    contempt for science and Scientifics, 19, 100-101
    Crimean War privations, 18, 45
    desertions, xiii, 12, 93, 101, 109, 131, 133, 177, 179, 192
    discipline, 18
    expenses, 21
    food thefts, 21, 65-66

"Hands to Bathe," 74
hardships and hazards of sea life, 15, 18, 19, 45, 50, 73-74, 85, 99, 119, 125, 179, 185-186, 231
    impressment, 18, 45
    literacy levels, 17-18
    shipboard life and routines, 45-46, 65-67, 74, 132, 148
    shore leave, 108-110, 177, 178, 192-193, 227, 231
Bonaparte, Napoleon, 127-128
Bonneycastle, Charles, 29-30
Botany Bay, 178
Brady, H. B., 249
Brazil, 121-126
    British enterprises in, 122-123
    local customs and culture, 123-124, 125
    Fernando Noronha penal colony, 118-120, 121
    vaqueiros and cattle industry, 124-125
Brewer, Peter, 83
British Admiral (steamer), 186
British Association for the Advancement of Science, 7
British Columbia, 188
Brooke, John Mercer, 28
Brooklyn Naval Yards, 234
Brown, Crum, 15
Bryozoans, 69, 71, 107
Buchanan, John Young
    career, 251
    death 251
    education and skills, 15-16
    expedition report, 249
    hardships, 21

laboratory on Challenger, 16, 48
and manganese nodules, 51-52, 54
sediment experiments, 63-64,
        133, 251
and von Willemoes Suhm, 16
Bumper subs, 241-242
Burgess Shale fauna, 188-189
Bush, Thomas, 231
Butkevich, V. S., 54

C

Cadrill I (drilling vessel), 244
Calcareous oozes, 135, 138
Calcite compensation depth, 61-65,
        84, 138
California Institute of Technology,
        172
Cambrian radiation, 188-189
Campbell, Lord George
    disgust with dredging, 61, 133
    on French colonialism, 227
    impressions and observations, 52,
            61, 111, 113, 114-115, 116,
            120, 121, 133, 134, 149,
            150-151, 192, 197, 198,
            199-200, 201-202, 218,
            220, 221, 224
    and Japanese culture, 218, 220,
            221
    Log Letters from Challenger, xiv,
            220, 251
    naval life as officer and aristocrat,
            92, 111, 118, 122, 148,
            218, 220, 230
    promotion and transfer, 230
    roots and personal characteristics,
            17
Canada, climate, 90

Canadian Journal of Physics, 225
Canary Current, 89, 103
Canary Islands, 51, 111
Cape Farewell, New Zealand, 185
Cape Finisterre, 248
Cape Hatteras, 89
Cape of Good Hope, 55, 92, 109,
        127, 129
Cape Town, South Africa, 67, 131,
        132
Cape Ushant, 248
Cape Verde Islands, 110, 111-112,
        214, 231
Cape York, Australia, 179, 193, 197
Carbon dioxide
    atmospheric, 162, 163
    hydrates, 82-83
    and subduction, 197
Carbon isotope ratios, 80, 81
Caribbean Sea, xiii
    agonic line, 75-76
    Bermuda Triangle, 74, 75-76, 78-
            79, 82, 84
    Blake Plateau, 78-79, 81, 82
    coral reefs, 69
    weather, 76, 91
Carlsberg Ridge, 60
Caroline Islands, xi, xiii
Carpenter, Lt., 202, 231
Carpenter, William, 2, 3, 4, 5, 20, 37,
        101, 134, 210, 217
Castle Harbor, Bermuda, 97
Caxoeira, Brazil, 123, 125
Cenozoic, 79, 157
Ceram (Seram) Island, 198
Ceratodus, 180, 182
Ceylon, 26
Chain (U.S. research ship), 38, 39
Challenger Deep, 204, 205, 212,
        215-217

Challenger expedition
 accidents and deaths, 15, 19, 50,
  73-74, 85-86, 99, 185-186,
  226, 231
 in Admiralty Islands, 203, 222
 approval, funding, and support, 5,
  180
 Antarctic leg, 157, 164-168
 Atlantic transect, 45-55, 60, 61-
  65, 111, 232, 236
 in Australia, 174-175, 176-179,
  180-181, 182-183, 184, 193
 in the Azores, 108-113, 231, 248
 in Brazil, 118-126
 burials at sea, 74, 226
 captain, officers, and crew, 12-13,
  17, 93, 111, 202-203, 230;
  see also Bluejackets
 correspondence from home, 91-
  92, 110
 crew's relationship with
  Scientifics, 17, 19, 21, 65-
  66, 93
 Darwin's influence, 5-6, 14, 16,
  34-35, 71, 117-118, 169,
  179-180, 228
 desertions of crew, xiii, 12, 93,
  101, 109, 131, 133, 177,
  179, 192
 discoveries and successes, 23, 27,
  28, 33, 34, 36-37, 51-55,
  60, 63-65, 84, 147, 204-
  205, 236, 252
 duration and extent, xiii, 230
 in Dutch East Indies, 198-201
 at Fernando Noronha penal
  colony, 118-120, 121
 fauna dredged up or discovered
  by, 27, 28, 34-35, 135, 136-
  137, 180-181, 186-187, 188

 food thefts, 21, 65-66
 in Friendly Isles and Fiji Islands,
  192-193
 at Gibraltar, 34-37
 in the Gulf Stream, 87, 93, 147
 at Halifax, Nova Scotia, 89, 90-
  93, 109
 at Hamilton, Bermuda, 84-86,
  94-99, 100
 in Hawaiian Archipelago, 222-224
 health issues and problems, 67,
  93, 94, 100, 108, 112, 125-
  126, 231, 252
 holidays and celebrations, 21, 94,
  147, 148-149, 229
 homeward bound, 229-231, 248
 in Hong Kong, 202-203, 220
 iceberg encounters, 154-155,
  166-167, 174-175, 178
 in Japan, 218-221
 at Kergeulen, 150-152
 lectures, 65
 legacy of, 172-173, 174, 252-254
 life insurance policies, 112-113
 at Lisbon, Portugal, 25-26
 manganese nodules, 51-55, 84,
  111, 147, 227, 252
 at Marion Island, 149-150, 151
 in New Zealand, 185-187, 192
 in Philippines, 201-202, 203
 published accounts of, 17
 purpose and objectives, 2, 4-6, 27,
  133
 report of Scientifics, xiii, 13, 135,
  249
 re-provisioning and refitting, 109,
  110, 112, 128, 129, 132,
  175, 178, 192
 in the Roaring Forties, 147, 148,
  154-155

route, x, xiii, 4, 101, 230
scientific staff ("Scientifics"), xiii–
    xiv, 13-16, 17, 20, 112-113
Sheerness to Portsmouth, 19-20
ship rescue mission, 73
shipboard life and routines, 17-
    18, 45-46, 48, 50, 65-66,
    92, 94, 101, 132, 148
shore leave, 108-110, 177, 192-
    193, 227, 231
sounding and dredging
    equipment and activities,
    10, 31, 33, 45, 46-49, 65,
    73, 87, 101. 110, 133, 148,
    166, 185-186, 204-205, 222
at South Africa, 130-133
at St. Paul's Rocks, 116-118
at St. Thomas, Virgin Islands, 65-
    69, 72-73
at Tahiti, 226, 227-228
telegraph cable surveying, 7, 178,
    185
and Termination Land, 164-168
at Tristan da Cunha, 127-129
at Valparaiso, 229-230
volcanic exploration, 200-201,
    222-223, 224
weather problems, 19-20, 21, 65,
    90-91, 112, 147-148, 150,
    155, 166-167, 185-186, 192
wildlife destruction, 149, 151
Challenger Lunar Module, 173, 253
Challenger Orbital Vehicle 99, 172-
    173, 253
Challenger Space Shuttle, 172-173, 253
Chambers Journal of Popular Literature,
    Science and Arts, 104
Charlotte Amalie, St. Thomas, 67-68
Chernobyl disaster, 163

China, 26, 189, 198
Chlorine hydrate, 77
Christ College, Cambridge, 251
Christmas Island, 13, 250
Circumpolar storm track, 153
Cita, Maria–Bianca, 41-42
Clathrates, 76-77. See also Carbon
    dioxide hydrates; Methane
    hydrates
CLIMAP (Climate: Long-Range
    Investigation, Mapping,
    and Prediction) project,
    144-146
Climate change
    Antarctica and, 154, 156, 158-159
    boundary conditions, 145-146
    glacial-interglacial cycles, 57, 82,
        95, 140-146, 158-159,
        160-162, 163, 164
    global warming, 80, 82, 83, 90, 156
    greenhouse gases and, 162, 163
    and Gulf Stream route, 90
    ice record, 158-159, 160-162,
        163-164
    Little Ice Age, 164
    methane hydrates and, 79-82
    modeling, 138-146
    sediment record, 138-146, 252
    sodium marker of storminess,
        163-164
    Younger Dryas, 163
Close Encounters of the Third Kind
    (film), 76
Cnidaria, 72
Cobalt, 53
Coccoliths, 107, 145
Coelacanths, 182
Coelenterazine system, 114
Cold War, 57-58, 214, 232-233, 239

Colladon, Jean Daniel, 29, 30
Colorado School of Mines, 78
Columbia University, 206, 207, 244
Columbus, Christopher, 87, 103,
105, 106
Commerell, Commodore, 130
Conan Doyle, Sir Arthur, 169
Connemara IV (ship), 103
Continental crust, 22, 23, 60, 196,
225, 239, 240
Continental drift theory, 38, 56-57,
58, 157-158, 196, 225, 246
Continental Oil, 240
Continental shelves, 21-23, 34, 58,
77, 84, 89
Continental slope, 29
Contuit Bay, Massachusetts, 207
Conway Morris, Simon, 189, 190
Cook, Captain James, 129-130, 148,
150, 151, 178, 227
Cook Strait, 185
Copper, 53
Coral reefs
atoll formation, 71
biological perspective, 71-72
cays, 68-69
Darwin's theory about, 71
diversity of species in, 69
fossil record, 189
framework builder, 70, 71-72
geographical distribution of, 69,
71-72
Great Barrier Reef, 73, 193
geological perspective, 70-71
nonsymbiotic, 70
photosymbiotic, 69, 71-72
polyp structure, 72
shipping hazards, 84, 186, 193
Coral Sea, 193
Corallium spp., 51

CORE (Consortium for Ocean
Research and
Exploration), 244
Cores and coring
climate record in, 140-146, 160-
162, 243
greenhouse gases in, 162
ice, 158-159, 160-164, 169-170
living fossils in, 164, 169-170
methane hydrates in sediment
samples, 78
pollution record in, 163-164
radioactivity in, 163
sediment, 140-146, 243-244
sodium marker of storminess in,
163-164
technology, 140, 142, 170
volcanic record in, 162-163
Cosmos Club, 238, 239
Cousteau, Jacques, 69
Crack in the World (film), 241
Cretaceous period, 157-158
Cretaceous-Tertiary boundary, 82
Crimean War, 18, 45
Crinoids, 27, 34, 118
Crozet Islands, 132, 146, 150, 151
CUSS I (drilling barge), 240, 241,
242, 245
Cyana submersible, 235
Cynodonts, 183

D

da Cunha, Tristão, 127
Dakar, Senegal, 213, 246
Dampier, William, 228
Darwin, Charles, xii, 2, 5-6, 13, 14,
16, 27, 34-35, 71, 117-118,
169, 179-180, 184, 191,
195, 228, 250

Dating
    with fossil record, 42, 246
    of methane hydrates, 79, 82
    radiometric, 225
    stalagmite, 96-97
Davy, Sir Humphrey, 77
de Gama, Vasco, 25-26
de Kerguelen-Tremarec, Yves Joseph,
    151
Deep ocean. *See also* Seawater
    calcite compensation depth, 61-
        65, 84, 138
    carbon dioxide storage as
        hydrates in, 82-83
    circulation, 4, 34, 79-80; *see also*
        specific currents
    dredging and sounding, 4, 10, 28-
        33, 45, 46-49, 65, 73, 87,
        101, 110, 133, 148, 166,
        185-186, 204-205, 222
    "evolutionary throwback" notion,
        28
    faunal populations, 210, 211-212,
        217
    gravels, 39
    gravitational fields in, 196
    hydrate formation in, 77-79, 82
    hydrothermal vents, 173, 235-236
    manned exploration, *see* Deep-sea
        exploration
    mid-ocean ridges, 29, 33, 53, 55-
        60, 62, 81, 84, 138, 195,
        196, 224, 231-236, 246
    pressure, 47, 77, 205, 206, 209,
        211
    sediments, *see* Sediments, seafloor
Deep Sea Drilling Project (DSDP),
        xii, 38, 42, 77-78, 144, 245,
        246, 253

Deep-sea drilling. *See also* Cores and
        coring; Deep Sea Drilling
        Project; *GLOMAR*
        Challenger
    bumper subs, 241-242
    dynamic positioning system, 241,
        245
    mobile oil platforms, 240
    Mohole project, 239-244, 245
    for sediment cores, 243-246
    vessels, 244
Deep-sea exploration
    in bathyscaphes, 205, 212-217,
        231, 234-235
    in bathyspheres, 206-209, 212
    benthoscope, 212, 214
    in Bermuda, 206, 209-212
    at Cape Verde Islands, 214
    Challenger Deep exploration,
        205, 215-217
    FAMOUS project, 231-236
    in Mediterranean Sea, 214
    of mid-Atlantic Ridge, 232-236
    radio broadcasts during, 211-212
    record dives, 209-210, 211, 212,
        214, 215-217
    robotic submersibles, 233
    sonar mapping, 235
    in submersibles, 232-236, 239
    thermocline layer and, 216
Deepwater formation, 64-65, 79-80,
        157-158
Defoe, Daniel, 13, 228-229
Deserts
    Antarctica as, 154
    seas and oceans as, 38-44, 107
Devonian period, 181
Diamond industry, 131-132
Dickens, Jerry, 81

Dickens, Mr., 219
Dinoflagellates, 114, 115, 137
Diploblastic body plan, 72
Disney, Walt, 181
Dipnoi, 181
Disraeli, Benjamin, 6
DNA sequencing, 169-170
Dolphin Rise, 29, 55
Douro (mail steamer), 252
Dredging. *See* Sounding and
    dredging
D'Urville Island, 185
Duck-billed platypus
    (*Ornithorhynchus paradoxus*),
    182-183, 184
Dutch East Indies, 198-201
Dwarf fauna, 40
Dysentery, 100

E

Earthquakes, 59, 81, 83, 192, 194,
    199, 202, 220, 235
East African Rift Valley, 58-59, 60,
    233
East Antarctic ice sheet, 170
East Islands, 150
Ebbels, Adam, 86, 99, 111
Echidna, 183, 184
Echo sounding, 29-33, 57, 78
Ecological niches, 27, 184
Elizabeth Island, 230
Ellen Austin (schooner), 102
Emancipation of slaves, 68
Emery, K. O., 232
Emiliani, Cesare, 142, 143, 161, 243-
    244
Emma Jane (whaler), 152
Emperor Seamount chain, 225

Endemic species, 97
Energy resources, 77
ENIAC (Electronic Numerical
    Integrator and Computer),
    239
Environmental stress, 40
Eocene epoch, 79, 158. *See also* P-E
    boundary
Equatorial current, 116
Ericson, David, 141-143
Erysipelas, 226
Euplectella subearea, 34-35, 136
Europa, 170-172, 173
European continental shelf, 21-22,
    34
Eutheria, 183
Evaporation, and current density, 89
Evaporite minerals, 39, 42
Evolution
    Cambrian radiation, 188-189
    chains of, 179-184
    Darwin's theory of descent with
        modification, 5, 16, 27,
        184, 195
    fossil record, 5, 27, 179-180
    missing links, 179-184, 186-191,
        252
    modern synthesis of, 252
    mosaic, 184
    natural selection, 5, 27, 184
    "throwback" notion, 28
Ewing, Maurice "Doc," 37, 57, 60,
    78, 141, 195, 232, 233,
    239, 244, 246
Exeter College, Oxford, 13, 250
Exocetus, 192
Extinction of species, 82, 99, 182,
    184
Extremophiles, 236

F

FAMOUS (French-American Mid-Ocean Undersea Studies) project, 231-236
Fantasia (film), 181
Farallon de Pajaros, 204
Farr, Harold, 32
Farraday, Michael, 77
Fathometers, 31, 57
Feira St. Anna, Brazil, 123-124
Fernando Noronha penal colony, 118-120, 121
Ferro Island, 51
Fessenden, Reginald A., 30
Fessenden Oscillator, 30
Field, A. J., 240
Fiji Islands, 10, 192-193
FNRS 2 bathyscaphe, 213
Fogs, 89, 156
Fonds Nationale de la Recherche Scientifique (FNRS), 213
Florida, 87-88
Florida Current, 88-89, 103
Fluorescence, 113
Flying fish, 74, 192
Foraminifera, 39-40, 42, 62, 80, 107, 134, 135, 136, 138, 140, 142, 145
Forbes, Edwin, 2, 3, 4, 62, 210, 217
Fossil record, 5, 27. *See also* Living fossils
    Burgess Shale fauna, 188-189
    of Cambrian radiation, 188-189
    of continental drift, 56, 57
    coral reefs, 189
    dating, 42, 246
    dwarf fauna, 40
    evolution in, 5, 27, 179-180
    lungfish, 182
    mass extinctions in, 82
    missing links in, 179-180, 182, 188-189
    stromatolites, 41
Fountain of youth, 87
Fox Dipping Circle, 34
France
    colonialism, 227
    deep-sea exploration, 213, 214, 231-236
Free, E.E., 208
Freshwater resources, 95, 154, 168-170
Friendly Isles, 192
Froelich, Paul, 32

G

Galileo spacecraft, 173
Ganymede, 173
Gastropod snails, 107
General Electric Company, 208
General Instruments Corporation of Massachusetts, 32
Genomanian-Turonian boundary, 82
Geological Society of America, 235
Geology
    of continental shelves, 22
    planetary, 22-24
Geothermal flux, 168
Germany, oceanographic research, 6
Gibraltar, 26, 31, 34-37
    Straits of, 40
Ginsburg, Bob, 243
Glaciations. *See* Ice ages and glaciations
Gladstone, William, 6
Glass, William, 128
Global warming, 80, 82, 83, 90, 156

Globigerina ooze, 51, 52, 61-62, 63, 64, 74, 133-134, 135
Globorotalia menardii, 140, 141
GLOMAR Challenger, 37, 38-39, 77-78, 245-247, 253
GLOMAR Explorer, 245
Global Marine Exploration Company, 240, 241, 244, 245
God versus Science, 7, 253
Good Words (magazine), 101
Gounong-Api volcano, 199
Gravels, 39
Gravitational fields, in deep ocean, 196
Gray ooze, 62
Great Barrier Reef, 73, 193
Great Britain
  in Africa, 112
  educational reforms, 17-18
  economic and maritime pre-eminence, 6, 26, 85, 123, 177
  foreign policy, 10
  Gulf Stream effects, 90
  moral code, 7
  navy, *see* Royal Navy
Great Southern Ice Barrier, xiii, 5, 93-94, 129, 134, 146, 159, 164-165
Green, Peter, 128
Greenhouse gases, 162, 163
Greenland, 90, 154
  ice cores, 161, 162, 163
  Viking colonies, 164
Greenland Sea
  deepwater formation, 64, 89
Grisley Folk, The (Wells), 44
Guadalupe island, 242
Guam, xi, xii, 204, 215

Guano/fertilizer industry, 13
Guatemala, 78
Gulf of Guinea, 112
Gulf of Maine, 232
Gulf of Mexico, 87, 106, 246
Gulf Stream, 147
  and climate, 76, 89, 90, 93
  discovery, 87-88
  features, 87, 88-89, 90
  mapping, 88
  and Sargasso Sea, 103, 106
Gutenberg, Beno, 59
Gypsum, 39
Gyres, 88, 138. *See also* Sargasso Sea

H

Haber, Fritz, 140
Haeckel, Ernst, 251
Haemocoel, 188
Half Mile Down (Beebe), 212
Halifax, Nova Scotia, 89, 90-93, 109
Hallucigenia, 189-190
Hamilton Harbor, Bermuda, 84-86, 94
Harrington Sound, 95, 97
Harris Anti-Submarine Warfare, 32
Harrow school, 13-14
Harston, Lt., 111
Hatchet fish, 210
Hawaiian Archipelago, xiii, 53, 173-174, 221, 222-226, 233
Hayes, Harvey C., 30-31, 57
Hayes Fathometer, 31, 57
Heard Island, 147, 153
Heezen, Bruce, 57-58, 59-60, 195, 225, 231
Heirtzler, Jim, 232
Hemiaster phillipi, 137

Hercules (ironclad), 35
Hersey, Brackett, 38, 39
Hess, Harry, 195-196, 237, 238, 240
Hilo Bay, 222
HMS *Agamemnon*, 18
HMS *Agincourt* (ironclad), 35
HMS *Audacious*, 18, 50
HMS *Beagle*, xii, 6, 14, 71, 117-118
HMS *Bounty*, 227
HMS *Captain*, 21
HMS *Challenger*. *See also Challenger*
        expedition
    builder, 7
    crews' quarters, 50
    design and layout, 1-2, 7-12
    foreign tours prior to expedition,
        10
    instrument and sampling
        platform, 11
    laboratories, 10, 12, 16, 47
    modern namesakes, 172-173, 253
    repairs, 178
HMS *Dido*, 178, 185
HMS *Duke*, 229
HMS *Essex*, 18
HMS *Invincible*, 18
HMS *Iron Duke*, 202
HMS *Lightning*, 3-4, 20, 22
HMS *Modesty*, 202
HMS *Pearl*, 185
HMS *Porcupine*, 4, 20, 22
HMS *Rattlesnake*, xii-xiii, 6, 130
HMS *Royal Alfred*, 93
HMS *Shearwater*, 4, 20, 22, 37
HMS *Simoon*, 112, 130
HMS *Sussex*, 18
HMS *Warrior*, 1, 85
Hobart, Tasmania, 176
Hog Island, 150
Home, David Milne, 96

Hong Kong, 202-203, 220
Honolulu, Hawaii, 222
Honor School of Natural Science,
        Oxford, 14
Horse Latitudes, 103
Horta, Pico island, 108
Hotspots
    biodiversity, 97
    volcanic, 224-226
Houot, Georges, 214
Hsu, Ken, 39-41, 42, 43
Hughes, Howard, 53
Humboldt Bay, 203
Humpbacked whales, 192
Huxley, Thomas Henry, xii-xiii, 4, 7,
        13, 15, 101, 130, 180, 226,
        251
Hydrates. *See* Carbon dioxide;
        Methane hydrates
Hydrothermal vent systems, 54, 173-
        174, 235-236

I

Ice ages and glaciations
    cause, 145
    glacial-interglacial cycles, 57, 82,
        95, 140-146, 158-159,
        160-162, 163, 164
    Little Ice Age, 164
    sediment record, 140-146, 243-244
    Younger Dryas, 163
Ice cores, 169-170
    climate record, 158-159, 160-164
Ice sheets, 22, 143, 156
    accumulation rates, 161
Icebergs, 154-156, 159, 166-167,
        173, 174-175, 178
Iceland, 60, 89-90

Illing, Vincent, 43
Ilo-Ilo island, 201
Imbrie, John, 144, 145
Imperial College (London), 43
Inaccessible Island, 127, 128
Indian Ocean, 131
    Carlsberg Ridge, 60
    circulation system, 88, 132
    coral reefs, 69
    earthquake epicenters, 59
Industrial Revolution, 163
Inouye, Admiral, 193
Inquisition, 26
Integrated Ocean Drilling Program,
    247, 253
International Date Line, 192
International Geophysical Year, 160
International Union of Geodesy and
    Geophysics, 239
Io, 171, 173
Ireland Island, 95
Iron, 53
Israel, 43

J

James B. Chester (bark), 102
Jameson, Robert, 2
Japan, 26, 53, 203, 218-221
Java, 197
Jellyfish, 115, 210
Jenkin, Fleeming, 3
John Murray (publisher), 250
JOI (Joint Oceanographic
    Institutions), 244
JOIDES (Joint Oceanographic
    Institutions for Deep
    Earth Sampling), xi, 244,
    245

JOIDES Resolution, xi, 253
Jet Propulsion Laboratory, 172
Juan Fernandez island, 228-229
Jupiter, moons of, 170, 171-172, 173

K

Kandavu, Fiji Islands, 192
Karakoa Ranges, 185
Karroo Desert, 131
Kauai, 225
Kenelm Chillingly (Lytton), 99
Kennedy, John F., 242
Kennett, James, 79-80
Kerguelen cabbage, 150, 151
Kerguelen (Desolation Island), 132,
    146, 150-152, 175
Ki island, 198
Kilauea crater, 223, 224
Kipp, Nilva, 144, 145
Krakatoa eruption, 197-198
Krefft, Johann, 181
Krummel, Otto, 104
Krupp Steel Works, 215
Kullenberg corer, 141
Kuroshio Current, 88
Kwajalein island, 71

L

Labrador Current, 89
Labrador Sea, 90
Lake Turkana, 58
Lake Vostok, 168-170, 173
Lamont-Doherty Geological
    Observatory, 37, 57, 232,
    244
Lancaster (U.S. corvette), 125
Langmuire, Irving, 105

Laurentia, 189
Le Pichon, Xavier, 232, 234
Lefroy, Governor, 94-95, 97
Lepidosiren, 182
LIBEC (LIght-BEhind-the-Camera)
        system, 233
Lidz, Louis, 143
Lill, Gordon, 238
Lisbon, Portugal, 25-26
Little Ice Age, 164
Lively (ironclad), 35
Living fossils, 98, 164, 169-170, 184
LOCO (LOng COres) committee,
        244
Lo'ihi Underwater Volcanic Vent
        Mission Probe, 173-174
Loihi volcano, 225-226
Log Letters from Chalenger (Campbell),
        251
Long Island, 185
Luciferin—luciferase system, 114,
        115
Lungfish, 181-182, 184, 191
Lusitania, 30
Lyell, Charles, 42, 43-44
Lyman, John, 244
Lytton, Lord, 99

M

M-reflector layer, 38-39, 40-41, 42
Madeira island, Azores, 35, 108, 110
Magma, rate of cooling of, 23-24
Majuro island, 71
Makian eruption, 199
Maldives, 26, 69
Malta, 37
Manganese nodules, 51-55, 84, 111,
        147, 227, 252

Manila, The Philippines, 201-202
Mariana Islands, 204-205
Mariana Trench, 215
Mariner mission, 171
Marion Island, 132, 147, 149-150,
        151, 175
Mars, 23
Marshall Islands, 71
Marsupials, 183
Mass extinctions, 82
Matkin, Charles, 94, 177, 220
Matkin, Joseph, 196
    in civil service, 252
    concerns about Captain
        Thomson, 202, 227-228
    death of, 252
    on deaths of fellow seamen, 186
    family, 17-18, 94, 110-111, 177,
        251
    father's illness and death, 21, 94,
        220-221
    on hardships of expedition, 21,
        50, 67, 86, 93, 179
    homesickness, 109, 229-230
    illness, 100
    impressions and observations, xiv,
        25, 26, 35, 73-74, 84, 91,
        108, 112, 116-117, 130,
        152, 166, 167, 192-194,
        198, 222, 227, 230
    marriage and family, 252
    personal characteristics, 18
    re-provisioning responsibility, 128
Matthews, Drummond, 246
Maud Rise, 79, 81
Mauna Loa, 222-223, 224
Mauritius, 69
Maury, Matthew, 29, 55
Mawson, Mr., 123

McDonald Island, 153
Mecom, John, 239-240
Mediterranean Sea
    basement topography, 40-41
    Chain survey, 38
    deep-sea drilling in, 214
    desiccation event, 38-44, 246
    M-reflector layer, 38-39, 40-41, 42
    Pillars of Atlantis, 41
    Shearwater expedition, 37
    trenches, 42-43
    Trieste test, 214
Melbourne, Australia, 176-177
Men o' war (ships), 1, 46, 84
Mercury (newspaper), 111
Mercury (planet), 171
Mesozoic era, 182, 183
Messier Channel, 230
Messinian salinity crisis, 42, 43-44
Messoyakha gas field, 77
Metameric segmentation, 187-188
Metatheria, 183
Meteor expedition, 57, 138, 140
Meteor impacts, 82
Methane, 81, 83-84, 162, 163, 170, 173
Methane hydrates
    and Bermuda Triangle myth, 76, 78-79, 82, 84
    as bottom simulating reflectors, 78
    cascade outgassing events, 81, 83-84
    and climate change, 79-82
    in core samples, 78
    dating of, 79, 82
    discovery, 77-78
    distribution of deposits, 77-79
    energy resource potential, 77, 81, 83-84
    formation process, 82
    stability, 77, 83
    structure and composition, 76-77
Mexico, 10, 242
Mid-Atlantic ridge, 53, 55-60, 62, 84, 98, 141, 147, 231-236
Mid-ocean ridges, 29, 33, 53, 55-60, 62, 81, 84, 138, 195, 196, 224, 231-236, 246
Middle island, 127
Milankovitch, Milutin, 145
Milne, Sir Alexander, 96, 97
Milne, Sir David, 96
Minotaur (ironclad), 35
Miocene Epoch, 42, 43-44
Missing links, 179-184, 186-191, 252
Mobile oil platforms, 240
Mohole project, 239-244, 245
Mohorovicic, Andre, 237
Mohorovicic Discontinuity, 237-238
Molucca passage, 198, 199
Moluccas, 26, 198, 249-250
Monitor, 85
Monotremes, 183-184, 191
Monteray Bay Research Institute, 83
Montevideo, 230, 231
Moon, 23, 24
Mosaic evolution, 184
Moseley, Henry Nottidge
    anthropological interests, 203, 219, 250
    boredom with dredging, xiii-xiv, 133
    death, 250
    education and scholarship, 13-14
    expedition report, 249, 250
    honors and awards, 250
    iceberg studies, 154-156, 159, 160, 161-162

impressions and observations,
121, 128-129, 154, 155
inspiration for natural history
interests, 13, 14, 228
laboratory on *Challenger*, 14, 16
land expeditions and experiences,
119, 121, 122-125, 182-
183, 200-201, 219
and missing links, 179, 182-183,
184, 186-187, 250
personal qualities, 14, 121, 250-
251
published works, 133, 250
scientific contributions, 250
volcanic studies, 200-201
and von Willlemoes Suhm, 16, 226
Mt. Toba eruption, 162-163
Multibeam sonar, 32-33
Murray, John
Brazilian interior expedition, 123
death, 250
expedition report, 249
guano fertilizer business, 13, 249-
250
interest in natural history, 13
leadership of Scientifics, 13, 249
scholarship, 13
sediment theory and cataloguing,
63, 133-134, 135, 146, 147,
252
Munk, Walter, 237, 238, 241
Museum of Natural History (New
York), 205-206
Mustard gas, 140

N

Napoleonic wars, 68, 127
Nares, Billy, 86, 111, 119, 120

Nares, George S., 12-13, 25, 26, 50,
67, 86, 90-91, 93, 97, 101,
108, 110, 111, 112-113,
118, 119, 125, 150, 165,
166-167, 168, 176, 177,
185, 186, 192, 202, 228
Nares Abyssal Plain, 103
Nares Harbor, Admiralty Islands, 203,
218
National Geographic, 58
National Academy of Sciences, 240,
242
National Aeronautics and Space
Administration (NASA),
170
National Geographic Society, 208,
210
National Science Foundation, 237,
240-241, 244, 245
Natural gas, 78
Natural selection, 5, 27, 184
Nature (journal), 101
Naval Electronics Research
Laboratory (U.S.), 215
Neoceratodus, 182
Neogloboquadrina pachyderma, 141
Neoproterozoic eon, 82
Neptune, 171, 174
New Foundland Grand Banks, 87
New Guinea, 203
New York, 90-91, 92
New York University, 208
New York Zoological Society, 206,
208
New Zealand, 178, 185-187, 192,
194
Nightingale Island, 127, 129
Nile River, 43
Nimitz, Admiral Chester William, 193
Nitrous oxides, 163

Nonsuch Island, Bermuda, 209
Norris, Richard, 81, 82
North America
    continental shelf, 58, 63, 89, 89
    Laurentia, 189
    methane hydrate deposits, 78
North Atlantic conveyor, 88, 89
North Atlantic Current, 103
North Atlantic Deep Water, 88, 89,
        90
North Atlantic Drift, 89
North Atlantic Ocean
    calcite compensation depth, 64
    circulation system, 88
    mid-ocean ridge, 60
    seismic profiles, 57-58
North Equatorial Drift, 103
North Pacific Ocean
    calcite compensation depth, 64
    mid-ocean ridge, 53
    faunal abundance, 222
Northern lights, 91
Norway, 3, 6, 89
Norwegian Sea, 89-90
Notes by a Naturalist on HMS
        Challenger (Moseley), 133,
        250
Nova Scotia, xiii, 89, 90-93, 109

O

Oahu, 222
Obi Island, 199
Ocean Drilling Program (ODP), xi-
        xii, 78, 79, 247, 253
Ocean temperatures, 4
    benthic isotope data, 144, 146
    and density, 89-90, 157, 158
    factor analysis method, 144

faunal abundance measurement
        methods, 141-143, 144
    Gulf Stream, 89
    hydrogen isotope method, 161,
        162, 163
    ice core record, 160-162
    ice volume effect, 143-144
    measuring, 46-48, 142-146, 205
    and methane hydrate stabi}Ÿty, 77
    oxygen isotope ratios and, 79, 80,
        142-143, 144, 145, 146,
        160-161, 162, 163, 243
    in Sargasso Sea, 103, 105, 107
    stability over glacial–interglacial
        cycle, 143-144
    thermocline layer, 216
Oceanic crust, 22, 23, 196-197, 204
Oceanic Hydrozoa (Huxley), 130
Oceans. See Deep Ocean; specific
        oceans and seas
Oil exploration, 70-71
Ontong Java Plateau, xi
Onycophora, 187-191
Origin of the Species by Means of
        Natural Selection (Darwin),
        71, 179, 180, 250
Orton, Arthur, 91-92
Osborn, Henry Fairfield, 206
Oxygen isotope ratios, 79, 80, 142-
        143, 144, 145, 146, 160-
        161, 162, 243
Oxyluciferin, 114

P

P-E boundary, 78, 81, 82, 83
Pacific Ocean
    agonic line, 75-76
    circulation system, 88

earthquake epicenters, 59
deepest part of, 204–205, 215–
216
mid-ocean ridge, 60
Ring of Fire, 194, 195–198, 222–
223, 230
Paleocene epoch, 79, 158. *See also* P–
E boundary
Paleothermometers, 79, 142–143
Paleozoic era, 182, 183
Pangea, 56
Papua New Guinea, 193–194
Pearl Harbor, 193
Penal colonies, 118–120
Penguins, 128–129
Peripatus, 187–191, 250
Permian period, 183
Permo-Triassic boundary, 182
Pernambuco, Brazil, 122
Persian Gulf, 69
Phanerozoic eon, 70, 188
Philippines, 34, 201–202, 203
*Philosophical Transactions of the Royal
Society of London* (journal),
77, 101, 134
Phoenix (German survey vessel), 15
Phosphorescence, 113
Photic excitation, 115
Photoproteins, 114
Photosynthesis, 69, 71–72, 80, 82
Piccard, Auguste, 212, 213, 214–215
Piccard, Jacques, 205, 213, 214, 215,
216
Pico island, 108
Pillars of Atlantis, 41
Pipefish (*Syngnathus pelagicus*), 106–
107
Pirates and piracy, 68
Pitt-Rivers anthropology collection,
250

Placental onycophorans, 191
Planetary crust, 22–23
Plankton, 140
Plate tectonics theory, 252
continental drift theory and, 38,
56–57, 58, 157, 196, 225,
246–247
hot spot theory, 224–226
directional change in Pacific
plate, 225
mid-ocean ridges and, 55–56,
232, 234
mid-plate volcanism and, 224–
226
processes in, 23
proof of, 42
rate of cooling of magma and,
23–24
seafloor spreading theory and,
195–197, 224–225, 246–
247
subduction zones, 23, 196, 204,
224
"Wound-that-never-heals
hypothesis and," 59–60
zone of crustal formation, 234,
235
Pleistocene, 95, 143, 243
Pliocene Epoch, 42, 43–44
Pollution record, 163–164
Polychaete worms, 107
Polymetallic nodules, 54
Ponce de Leon, Don Juan, 87–88
Port Hardy, D'Urville Island, 185
Port Moresby, New Guinea, 193
Port Nicholson, 186
Port Stanley, Falkland Islands, 230
Portugal and Portuguese people, 25–
26
Portsmouth, England, in 1872, 1–2

Possession Island, 150
Power, Albert E., 105
Precambrian–Cambrian boundary, 188
Pressure, deep-ocean, 47, 77, 205, 206, 209, 211
Primary crust, 22-23
Prince Edward Island, 132, 150
Princeton University, 195
Proterozoic eon, 70, 188
Protopterus, 182
Prototheria, 183
Pteropod ooze, 61-62, 63, 133
Puerto Rico Trench, 196

Q

Queen Charlotte Sound, 185

R

Rabi, I. I., 240
Radioactivity, 163
Radiolaria, 135, 136, 145
Raine Island, 193
Ready (research ship), 208, 209, 211
Red clay sediments, 55, 61, 62, 64, 133
Red Sea, 58, 60
    coral reefs, 69
Rhone River, 43
Richter, Charles, 59
Rift valleys, 58-60, 233
Ring of Fire, 194, 195-198, 222-223, 230
Roaring Forties, 127, 131, 132-133, 147, 148, 153, 154-155
Robinson Crusoe (Defoe), 13, 228-229
Robot sampling, 170, 173-174, 233

Rock cycle, 23-24
Rogers, Woodes, 229
Rolleston, George, 14
Roosevelt, Theodore, 206-207
Rosalie (merchant ship), 102
Ross ice shelf, 154
Roswell King (whaler), 152
Royal Naval Shipyards at Woolwich, 7
Royal Navy, 1-2. *See also* Bluejackets
    Channel Fleet, 35
    customs and traditions, 118
    discipline, 227
    men o' war, 1, 46, 85
    North American Fleet, 85
    turret warship, 85
Royal Society of Edinburgh, 96
Royal Society of London, 3, 4, 101, 240, 250
Ryan, Bill, 38-41

S

Sabhka, 41
Sagan, Carl, 172
Salt domes, 246
Saltpans Drift, 131
San Miguel, Azores, 108-109
Sandwich Islands, 53, 221, 227. *See also* Hawaiian Archipelago
Santa Cruz, 50
Sarcopterygii, 181
Sargasso Sea, 147
    boundary currents, 103, 104, 107
    as a desert, 107
    ecology, 103, 105-107
    legends of lost ships and crews, 102-103, 104
    location and area, 103, 104
    ocean floor beneath, 103-104

pollution problems, 107–108
thermal structure, 103, 105, 107
winds, 103, 105
Sargassum kelp, 103, 105–106
Sarmiento Channel, 230
SASS (Sing-Around-Sonar-System), 232-233
Sassen, Roger, 84
Saturn, 171, 174
Scandinavia, oceanographic research, 6
Schmitz, Birger, 81
Schott, Wolfgang, 140, 141
Science (journal), 145
Scientific American (magazine), 241
Scotland, 90
Scott, R. F., 165
Scripps Institution of Oceanography, 18, 215, 237, 242, 244, 245, 251
Scurvy, 94, 112, 151
Sea-Beam sonar, 33
Sea level, ice sheets and, 22, 95
Sea lily, 27, 28
Sea-spiders, 107
Seafloor spreading, 195–196, 224–225, 231, 233, 246-246
Seawater
  acidity, 55, 60, 63–65, 138
  density changes, 76, 79, 80, 89, 156-157
  on Europa, 173
  gold extraction from, 140
Secondary crust, 23
Sediments, seafloor. *See also specific types of sediment*
  barium in, 81
  classification of, 135, 252
  climate modeling from, 138–146, 243-244, 252

cores and coring technology, 140-146, 243-246
depth and, 61–65
geological timescale, 138–139
at mid-ocean ridges, 138, 246-247
oxygen isotope ratios, 79
sampling apparatus, 46, 48
sources of, 133–137
thickness and age of, 135, 138
Seed-shrimp (*Vargula*), 114
Seismic shooting, 37–38
Seismic wave velocity
  of bottom simulating reflectors, 78
  Mohorovicic Discontinuity and, 237-238
Seismic wave mapping, 59, 233
Selkirk, Alexander, 228, 229
Severn Bore, 95
Seychelles, 69
Shackleton, Ernest, 165
Shackleton, Nick, 143–144
Shark Bay (Western Australia), 41
Sharks, 65
Shell Oil, 240
Shepard, Alan, 242
Shipwrecks and shipping disasters, 29, 30, 73, 84, 86, 91, 92, 102-103, 128, 186, 193
Siberia, 60, 77
Siberian Tundra, 77
Sierra Leone, 112
Siliceous oozes, 135, 138
Simon's Bay, 130
Simonstown, South Africa, 67, 109, 129, 130-133
Six, James, 46
Slavery, 68
Smallpox epidemics, 108, 110
Smithsonian Museum, 188

Society Islands, 226

Sodium, as marker of storminess, 163-164

Sombrero, Virgin Islands, 50

Sonar (SOund NAvigation and Ranging), 31, 32-33, 38, 232-233, 235, 241

Sounding and dredging
Challenger's equipment and activities, 10, 31, 33, 45, 46-49, 65, 73, 87, 101, 110, 133, 148, 166, 185-186, 204-205, 222
conventional, 46
principle, 28
techniques and technology, 28-33, 46, 57, 58, 235

South Africa, xiii, 67, 133
Ashanti Wars, 112, 130-131
diamond industry, 131-132
missing links, 187

South America, missing links, 182, 187

South Atlantic Ocean, mid-ocean ridge, 60

South Kensington Museum, 15

South Pacific Ocean, mid-ocean ridges, 53

Southern Ocean, 93-94
chemical and biological distinctiveness, 153
currents, 154, 156, 158
islands, 149
mixing of water masses, 156-157
sediments, 62
temperature and salinity, 152, 156-157

South Pole, 160, 165

Speed of sound in seawater, 29, 30, 31

Spice Islands, 198

Spielberg, Steven, 76

Spiny anteater, 183

Spry, William J.
boredom with sounding and dredging, 133
colonist prejudices of, 25, 35, 36
Cruise of the HMS Challenger, xiv, 17, 251
on death of seaman, 74, 85
engineering skills and duties, 46, 67, 109, 121
impressions and observations, 20-21, 36, 67, 85, 90, 125, 128, 131, 151, 167, 178, 203, 228, 248
interest in missing links, 179, 180-181
naval life and privileges, 125, 132, 230-231
and Scientifics, 48, 157, 179, 180-181

Sputnik, 214, 237

St. Amaro, Brazil, 125

St. Helena island, 127-128

St. Iago (Santiago) island, 112

St. Paul's Rocks, 113, 115, 116-118, 121, 134

St. Thomas, Virgin Islands, 60, 65-69, 72-73

St. Vincent island, Azores, 111-112

Star of South Africa (diamond), 131

Stokes, William, 73-74, 86, 99

Stoltenhoff, Gustav and Frederick, 129

Stoltenhoff island, 127

Stott, Lowell, 79-80

Stradling, Captain, 228-229

Strait of Magellan, 230

Stromatolites, 41

Sub-Antarctic intermediate water, 156
Sub-Antarctic surface water, 153
Subduction zones, 196, 197, 204, 224
Submarine Signal Company, 30
Submersibles, 232-236, 239 . *See also* Deep-sea exploration; *individual submersibles*
Subtropical Convergence, 153, 152
Sultan (ironclad), 35
Sumatra, 26, 197
Sunda, 197
Superficial segmentation, 189
Superior Oil, 240
Surtsey, 233
Swift, Don, 22
Swift, Mary, 252
Swire, Herbert
   colonialist prejudices, 119
   death, 252
   illness, 231, 252
   impressions and observations, 69, 100, 111, 120, 201, 223
   marriage and family, 252
   Moseley's friendship with, 250-251
   naval life, 26, 34, 35, 65, 110, 120, 123, 125, 126, 174-175, 201, 222
   personal characteristics, 17, 99, 175, 231
   relationship with Scientifics, xiv, 17, 93, 99, 110, 123, 201, 204
   and women, 201, 219-220
Swire Deep, 204-205
Sydney, Australia, 177-178
Sydney Heads, 177
Syria, 43

T

Tahiti, 53, 226, 227-228, 231
Taylor, Alfred, 91, 128
Telegraph cable, underwater, 7, 37, 178, 185
Temperatures. *See also* Ocean temperatures
   Antarctic, 160
   P-E boundary, 79
Tenerife, 50
Termination Land, 164-168, 174. *See also* Antarctica
Ternate Island, 199-201
Terra Australis Incognita, 165-166
Tertiary crust, 23
Tertiary era, 157
Test Ban Treaty, 163
Tethys Ocean, 38, 158
Texas A&M University, 84
Tharp, Marie, 57, 58-59, 195, 225, 231
The Deep and the Past (Ericson and Wollin), 143
Therapsids, 183
Thermal isolation, 158
Thermocline layer, 216
Thermometers
   deep-sea, 46-48, 49
   paleothermometers, 79, 142
Thomson, Frank Thurle, 202, 227-228, 229
Tichborne, Sir Roger, 92
Tichborne Claimant trial, 91-92
Tidore island, 199
Tierra del Fuego, xiii, 230
Titanic, 29, 30, 86
Titanium, 53
Tizard, T. H., 116, 119, 249
Tonga, Friendly Isles, 192

Topography of ocean floor
    Atlantic Ocean, 29
    echo sounding maps, 57
    first chart, 29
    continental shelves, 22
    hand-drawn perspective maps,
        57-58, 195
    mid-Atlantic Ridge, 232-233
    Ontong Java Plateau, xi
    ridge and swale, 22
    seismic wave mapping, 59, 233
    technology development for
        mapping, 28-33, 232-234
Torres Straits, 193-194, 196, 197
Treaty of Fomena, 130-131
Treaty of Utrecht, 35
Treaty of Versailles, 140
Tres Montes peninsula, 230
Triassic period, 183
Trieste bathyscaphes, 205, 213-217,
        231, 232, 235, 239
Tristan da Cunha island, 126, 127-
        130, 152
Trolltunga iceberg, 156
Truman, Harry, 238
Tubbs, Tom, 19
Tubeworms, 236
Tufts University, 206
Tyrenhian Sea, 214-215

U

Umbellularia, 118
Underwater bells, 30
Union Oil, 240
United States, oceanographic
        research, 6
University of Bonn, 14

University of California at Santa
        Barbara, 79
University of Cambridge, 143
University of Edinburgh, 2-3, 4, 13,
        15
    Natural History Museum, 96
University of Iowa, 57
University of Miami, 243, 244
University of Munich, 15
University of Oxford, xiii, 250
University of Toronto, 224
University of Virginia, 29
University of Washington, 244
Upwelling zones, 138, 157
Uranus, 171, 174
Urey, Harold, 142, 243
U.S. Naval Applied Science
        Laboratory, 234
U.S. Naval Research Laboratory, 233
U.S. Navy
    bathymetric contour maps, 57-58
    bathyscaphe, 214-216
    Depot of Charts and Instruments,
        29
    Gulf Stream mapping, 88
    Office of Naval Research, 214-
        215, 238-239
    sounding technology, 29-33
    World War II, 193
USS *Lexington*, 193
USS *Stewart*, 31, 57
USS Wandank, 215-216
USS *Yorktown*, 193

V

Valparaiso, 53, 229-230
Vandenberg Air Force Base, 71
Varuna (British ship), 73

Venus, 23, 24
Venus Flower Basket, 34
Verne, Jules, xi
Victoria Falls, 40
Vigo, Portugal, 248
Vine, Fred, 246
Virgin Islands, 50, 60, 65-69, 72-73
Volcanoes and volcanism
    atoll formation, 71
    and extremophiles, 236
    hot spot theory, 224-226
    hotel and bathing contraptions at
        edge of, 223-224
    in Hawaiian Archipelago, 222-
        223, 224, 225-226
    and hydrate stability, 83
    hydrothermal vent systems, 54,
        173-174, 235-236
    ice core record of eruptions, 162-
        163
    on Jupiter's moons, 171-172
    Krakatoa eruption, 197-198
    Makian eruption, 199
    mid-plate activity, 224-226
    Ring of Fire, 194, 195-198, 222-
        223, 230
    in Spice Islands, 198-199
    at subduction zones, 196-197,
        204
    submarine, 71, 83, 173-174, 225-
        226, 235-236
    vents, 223, 235-236
von Willemoes Suhm, Rudolf, 14-
        15, 16, 93, 226, 249
Vostok Ice Station, 160-162, 168-
        169
Voyage of the Beagle (Darwin), 14
Voyager missions, 170-172, 174

W

Walcott, Charles, 188
Walcott's Quarry, 188, 189
Walsh, Don, 205, 215, 216
Walsingham caves
    Admiral's Cave, 96
    Bassett's Cave, 98
    ecological threats to, 98-99
    fauna, 97-98
    formation, 95-96
    Painter's Vale Cave, 97
    stalagmite dating, 96-97
Water sampling flasks, 46, 47, 49
Watson-Stillman Company, 208
Wave interference patterns, 32
Washington (U.S. Navy brig), 29-30
Water temperatures. See Ocean
        temperatures
Weather. See also Climate change
    Antarctic Bottom Water and, 154
    in Bermuda triangle, 75-76
    sodium marker of storminess,
        163-164
Wegener, Alfred, 56-57, 58, 196, 225
Wellington, New Zealand, 186, 187
Wells, Herbert George, 43-44
West Indies, 55, 65-69, 72-73, 106
West wind drift, 153
Western boundary currents, 88
Whalers and whaling, 152-153
"White Man's Grave," 67, 112
White Star Line, 86
Wilberforce, Bishop of Oxford, 7
Wild, John James, 16, 135
Wilkes, Captain, 165-166, 167
Willm, Pierre, 214
Wilson, J. Tuzo, 224-225, 226
Wing, Otto, 104
Winton, Edward, 185-186

Wollin, Goesta, 141-143
Wolseley, Garnet, 130
Woods Hole Oceanographic
		Institution, 38, 81, 141,
		232, 233, 234, 244
World War I, 140
World War II, 193, 203, 212, 213, 238
Worzel, Joe, 246
"Wound-that-never-heals
		hypothesis," 59-60
Wurm ice shelf, 154
Wyatt, Richard, 86
Wyville Thomson, Charles, 51
	and calcite compensation depth,
		63-64
	Carpenter's relationship with, 3-
		4, 5, 20, 101, 134
	and Conan Doyle's Professor
		Challenger, 169
	death, 249
	deepwater circulation theory, 34
	discoveries, 4, 180-181, 210, 217
	family, 20
	and King of Portugal, 26
	impressions and observations, 10,
		117, 118-119, 128, 151,
		165, 168
	Lightning expedition, 3-4
	published articles, 53, 77, 101-
		102, 134, 147, 166

	at Fernando Noronha penal
		colony, 118-120
	and manganese nodules, 53, 54
	Nares' relationship with, 202
	Porcupine expedition, 4
	relationship with other
		Scientifics, 15, 118, 226
	scientific interests, 61
	sediment theory, 133-134, 146,
		147, 252
	Shearwater expedition, 4, 20, 37
	at University of Edinburgh, 2, 4,
		96
	at Walsingham Caves, 94-95, 96-
		97

Y

Yale University, 105
Yarra River, 183
Yellow fever, 67, 125-126
Younger Dryas, 163

Z

Zamboanga, The Philippines, 201
Zone of crustal formation, 234
Zooanthellae, 71, 72
Zoutkloof, South Africa, 131

Cast of Characters
Picture Acknowledgements

Henry Moseley: Alphabet and Image.

William J. J. Spry: Getty Images–Time Life.

Officers group: From Prof. Moseley's Albums, Dept of Zoology, Univ. of Oxford.

Captain George S. Nares: National Portrait Gallery, London.

John J. Wild: Self-portrait from *At Anchor*. Photo: Michael Holford.

John Murray: From Prof. Moseley's Albums, Dept of Zoology, Univ. of Oxford.